# MY COUNTDOWN
### The Story Behind My Husband's Spaceflight

An Apogee Prime Publication

**Brenda, Julia, Bob, Roman, Frank,**
                          **thank you for trusting me**

## Six, five, four, three, two, one – LIFT-OFF!

With a blast of spectacular flames The Rocket carrying my husband and his two good friends shot straight into a flawlessly blue sky. It was a perfect day and a perfect launch. The culmination of dreams, aspirations, years and years of training. Six months to go until we are back together.

We have a LIFT-OFF – and this is when the real countdown began for me.

**Your job is only to write your heart out, and let destiny take care of the rest.**

*Elizabeth Gilbert*

# Contents

| | |
|---|---|
| FOREWORD | 13 |
| INTRODUCTION (HOW IT ALL STARTED) | 15 |
| 6 THE ISS PROGRAMME IN A NUTSHELL | 25 |
| How is it organized? | 25 |
| What does it mean to train for a long-duration flight to the ISS? | 33 |
| Star City, Moscow, Russia | 33 |
| Visiting the Kremlin and Red Square | 35 |
| Useful numerical superstition | 39 |
| Big dogs | 41 |
| And this has also happened | 44 |
| 'The Planets' | 45 |
| Houston, USA | 45 |
| Chris, the Dog, the First Bearer of the Charm | 46 |
| Big astronaut family | 47 |
| Montreal, Canada | 50 |
| Kyoto, Tokyo, Tsukuba, Japan | 53 |
| Cologne, Europe | 57 |
| 5 BAIKONUR | 59 |
| 4 THE LAUNCH ACCORDING TO... | 71 |
| ...Roman | 71 |
| ...Julia | 74 |
| ...Bob | 76 |
| ...Brenda | 80 |
| ...Frank | 82 |
| ...Lena | 84 |

## 3 The docking according to... 91

...Bob 91
...Frank 93
...Roman 94
...Brenda 96
...Lena 98
...Julia 104

## 2 Personal life during the flight, in space, on the ground 107

There is none! 107
Divine inspiration 110
Like Captain Jean-Luc Picard 114
Sunny side up 118
Multi-Culti 119
Multi-Culti 2 121
Multi-Culti 3 122
My best commander in the world 123
Surreal but true 124
One other thing you only know later 127
Staying healthy, staying in touch 128
Fifty fifty 130
To pee or not to pee 132
'The Bet' 135
Thinking positive might help 136
The third quarter 137
8 September 143
Halloween 2009 143
The world lasts because it laughs 145
Save and protect 146

## 1 Landing — 149

| | |
|---|---:|
| Me | 149 |
| Him | 153 |
| Us | 155 |
| Vstrecha | 156 |
| And life goes on | 159 |

## A few closing words — 161

| | |
|---|---:|
| Space, microscopes and future cosmonauts | 161 |
| A shooting star | 162 |
| To all those good people | 164 |
| To be continued | 168 |
| About UNICEF | 170 |

## Opruimte

Eén keer per jaar
ruim ik de ruimte op
plaats ik alles precies juist

Maar maakt de aarde
een pirouette
dan dansen alle sterren mee

Ik ruimte
Ik ruimde
volgens de regels

*Nele De Winne*
*(Frank's daughter)*

## I Tidy Space

Once a year
I tidy up Space
Place everything in place
But when the Earth spins
All the stars dance along

I tidied up space
I'm tied up in space
I found my place

GENNADI PADALKA

# Foreword

As I was departing for my third long-duration spaceflight, I would have never guessed that it's possible to surprise me anymore. I know firsthand this serious work and the strict discipline, professional camaraderie and space brotherhood. I know which magnificent heavenly feelings and emotional hardships are experienced by the crews in space and their loved ones on Earth.

Frank, Roman and Bob joined the Expedition 20 on the International Space Station at the end of May. Upon their arrival, for the first time we became six people on board. Our space brotherhood has grown stronger: we were an exceptionally connected group. Once, at the beginning of July, Frank asked me if I would be interested to read what his wife Lena was writing about their flight. For me it was a pleasant surprise that they found my opinion important. I started reading and I could not stop, I was looking forward to the following chapters for the remainder of my flight. Through the prism of Lena's story, I saw a totally new side of life of all our wives; their worries, their concerns but most importantly their infinite love and care for us.

You have in front of you a very genuine, very honest story. This story is filled with authentic emotions that are set out in very precise technical context. This is a monologue of a woman. She is loving, understanding, compassionate and supportive of her man. The man is surrounded by a fragile string of constellations and planets beyond which there is nothing but black nothingness. This is a confession of a woman who knows that there is no life without troubles and no love without worries.

Gennadi Padalka
Cosmonaut, Colonel
Hero of the Russian Federation
Soyuz TMA-14 Commander
ISS Expedition 20 Commander

# Introduction (how it all started)

A weekend is one of the two days when I get to see Frank for over 30 minutes, floating around in one of the modules he chooses for holding our family conference.

But more importantly, Frank gets to see me. When I want, I can look into the NASA TV web-site – everybody can. This way I can see what the guys are doing. But they don't have a chance to see anything unless it's specifically sent or broadcast to them. I have video conference equipment at home. Before every conference, I try to think what would be nice for Frank to see for a change. No pun intended, but life on board is very limited in terms of space. I take the monitor into different rooms in our house. I show Frank different views from the windows. Our neighbours on the right have replaced the garden fence. I bet they have no idea that images of their new fence and a glimpse of their new garden umbrella and patio tiles were broadcast to space!

When I run out of rooms, and new views, where the cables from the conference monitor will reach (this is bound to happen by the end of July), I might hold one of my family conferences from the bathroom. Frank will be so jealous – going to the toilet on board is a really long affair, resulting in cleaning and brushing of various utilities every time. Hopefully the Houston flight controllers wouldn't mind. I had a chance to meet briefly before the flight with a few people who support the family conferences. They assured me that they have no opinion about anything they might accidentally see (they have to look occasionally to be sure the link is working).

Frank calls me on the phone at least a couple of times per day. It's like he's on a very long business trip, but one that I couldn't join him on this time. The phone on the International Space Station (ISS) adds a sense of normality into my life. Frank always calls when he has a chance when he travels. The onboard phone is the most private way of communication from space. In our digital day and age, when anything is no longer really private, this is at least as good as any terrestrial phone. I can't tell you how much I feel for those astronauts and their families who flew many years ago: infrequent contact with Mission Control, and no privacy for talking for the whole duration of the flight.

Video conferences at the weekends make Frank's current 'business destination' – the Space Station – more real in my life. Every Sunday, I can't help facing the fact that my husband and his friends are sitting (well, floating) in a sophisticated tin in airless space and their lives depend completely on a large group of very clever and well-organized people spread around the globe. I have to trust those people to bring my husband back safely and I choose to do so. Nothing would change if I didn't trust them, except only that I would feel much worse.

When my mind starts wandering off into the expected and too predictable feminine escapades of whimsical anxiety, I just remind myself that we (me, my mind, my body and spirit) agreed to deal with this particular business trip in the same way we deal with any other business trip – we assume that everything will be fine and look forward to Frank coming back; fantasizing about what to eat for the first evening of his return, what to wear, and planning when to see the rest of the family. After all, it's his job. And since he's working up there, this is just a business trip, right?

I feel bad for a fraction of a second when the conference ends and the screen goes black. I don't like goodbyes and separations, I never did. It brings tears to my eyes, but I know by now how to stop them halfway and put the pent-up flow of sadness into reverse. I had a lot of time to practice in anticipation of the launch. Their departure from Star City to Baikonur a fortnight before the launch was the toughest. Maybe this was because it was the first real separation in a series of controlled separations in the two weeks and six months to come?

Or was it because I wasn't allowed to travel with Frank to spend a few more days together before the launch? The latter was a more probable reason. I'll tell you more about this later. But now everything is fine. Frank will call on the phone in a couple of hours anyway. He always does when he's away from home – on a business trip.

* * *

It was Sunday, 28 June 2009. This time Frank chose the Japanese 'Kibo' module for holding the family conference on his side. 'Kibo' means 'hope' in Japanese. Most of the modules have technical titles and names. Kibo is a marvel of Japanese technology. It had arrived at The Station quite recently. The wall behind Frank looked almost as pristine as its original prelaunch pictures.

About 20 minutes into our discussion, Bob (Thirsk) flew by and waved at me. I still can't get over how cool, but at the same time normal, this sounds to me these days – that somebody just floated or flew by. And waved at me personally.

My conversation with Frank at this point was actually going very well for me. I was negotiating around the fact that, since I'm living alone for six months, I have to take care of rolling out the garbage containers myself. They are heavy and I have just broken a nail again, damaging the finger tissue while I was doing it. Frank is great at promising me things that he knows will make me feel better. Last week, I had to deal with the annual car service and yesterday with a crashed internet at home. These facts were contributing enormously to the strength of my position going into the negotiation. I was laughingly rejoicing because we had just reached a firm agreement that he would take out the garbage within 10 minutes of only one reminder. Family conferences are recorded as a personal souvenir for the astronauts and their families. For the first time in my life, I will have official evidence that my husband agrees that it's a man's chore to deal with the garbage and an unambiguous commitment to do so.

Ha! Dear men reading this, I am not a nag; we (women) all think like this. It's just that some choose to be more diplomatic and give up asking for help. Frank tried to appeal to my sense of compassion: he also had to deal with the garbage this week. All the trash has to be put together for disposal by a departing Progress cargo spacecraft. Each Progress burns up on reentry into Earth's atmosphere. This is one of the methods of garbage disposal for the Space Station. I hesitated for a fraction of a second: I'm inclined to be instantly supportive when someone complains about a problem in their life (a typical female reaction, related to stimulation of serotonin production in the brain). But I very quickly came back to my senses – this was all part of the business trip. You don't go to a scout camp and expect to have a bath-tub in the middle of a forest. You don't fly to space and expect that the household is taken care of by someone else.

Bob smiled and waved again. He knew about my ongoing struggle – whether I should stay in my nice secure corporate job, or should I let go of it and take a full blast at writing. I started writing fiction stories for a glossy Russian magazine at the end of last year. Even though it was my first real writing experience (press-releases and reports don't count), to my great surprise the magazine loved my stories and asked me to keep writing. But it was not a job. Those were just two-page monthlies.

Bob was still waving and smiling on the screen. He heard earlier today that Frank and I were again discussing what should I do.

"I know it can be a tough decision to make, Lena," said Bob. "We've been talking about this a lot with Frank. You've got to follow your heart. It's not always easy."

"I know. At least it's nice for me to look at you guys. You always look so happy with what you are doing," I said.

"Yes we are. We're having a lot of fun up here," said Bob. "Too bad you're not here with us. You could write about our life. There's so much happening in our lives that has nothing to do with our technical work. It's a shame to see that it'll pass by unknown to anybody."

"How do you mean?" I asked.

"This is a spectacular human experience to live with six people. This would be so interesting for everybody to know. Too bad it's not recorded in any way. You've been through a lot already with us in the last years. If only you could be the seventh up here. I know you're not very technical, but there'd be so much for you to do, just writing about what's going on. Why don't you write about it anyway?"

"That sounds very interesting. How is Brenda?"

"I'll talk to her later today, I'll know more but I think everything is fine. I called Lisane recently (Bob's daughter). She's currently in Mexico. Anyway, I'll leave you two to your talk, see you soon," said Bob.

"Give my love to Brenda when you see her."

"Will do," said Bob. And he slowly floated away.

I spent most of the rest of the day writing, and sent the file to Frank and went to sleep.

It was 28 June, 2009. One month and one day into the flight. No, we are not counting days. But it feels kind of nice that there are less than five months to go.

* * *

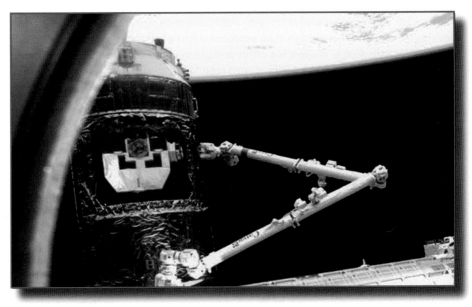

JAPANESE MODULE SEEN THROUGH THE STATION WINDOW

Frank had always been extremely particular about 'no personal life' in his professional appearances. I generally agree with him.

The profession of an astronaut seems to be perceived by people in a way similar to being a Hollywood actor. Not quite a superstar, but a celebrity nevertheless. Most people don't think about that when a person chooses a performance career, whether to be an actor, singer, artist or a writer (yes, I'm fully aware of the contradiction in me writing this book) or other job where the product of their work is aimed directly at the public, they know that their professional success depends on the acceptance of their art by the public. You can only be a successful artist, in the broadest meaning of this word, if people want the results of your work – as simple as that.

But how many of you know any Nobel Prize winners in physics or chemistry? I certainly don't. Those people have done no less for global society than the world-famous actors and singers. Do you remember the name of the captain who made that emergency landing of his plane on the Hudson River at the beginning of 2009? I don't. Why go even that far, if we are talking about space. The ISS programme is possible because of the vision and will of various major global players to make it work. Do you know the names of the directors of the agencies that participate in the ISS? Do you know the names of the politicians who have signed the intergovernmental agreements that govern the ISS programme? Those are the people who shape the environment that makes human spaceflight possible.

You could argue that, like in a good movie, the script writers, the camera operators and the directors and so on always stay behind the scenes, while the actors are those who get the real fame and recognition. Yes. The difference is that acting is a trade that is built on having a talent aimed at drawing public attention, mesmerizing the audiences, leaving them in awe. But the job of an astronaut is first of all about technical excellence in what they are doing. While a big effort is made to communicate the ins and outs of the profession to everyone who is interested (and to interest those who aren't), the core of the job is about doing everything impeccably right, and not being spectacular in it.

Astronauts, by the nature of their jobs, are like the Nobel Prize winners in sciences. The content of their jobs greatly contributes to the betterment of society. If you look into this matter from a purely technical perspective, the work they're doing might be as complicated. But at the same time, the nature of their deeds resonates so much with human nature: the exploration of the unknown. The public image of their work often crosses over and gets romanticized in a 'Hollywood' perception of what they're doing. And of course it would be naïve to neglect the aspect of the desire to have something to do with famous people.

I was born and grew up in Moscow. My life circumstances turned out in such a way that, in 1992, I pretty much woke up one day in Holland. When people ask me where I'm from, my usual answer is, "I'm born Russian, but technically I'm Dutch." We once stopped to buy food in a Russian shop in Brussels that we'd seen as we drove by. Frank and I were exchanging some comments in a mixture of Russian, English and Dutch as we randomly wandered through the shop. After a few minutes, the woman running the shop turned to me.

"How come you are shopping with our astronaut?" she said in Russian, with a smile but with a rather firm demand for an answer. I get taken aback by direct inquiries. I wasn't sure what to say.

"Actually, I'm shopping with my husband," I answered with the most intelligent thing I could think of.

"It can't be that our astronaut has a Russian wife," she blurted as her face noticeably changed expression in disbelief.

"No, he doesn't, he has a Dutch wife," I chickened out of an unnecessary conversation.

"Oh, you speak such good Russian," and now it was her turn to be taken aback.

I agreed, smiled and helped her to arrange for her friend's child to have a photo taken with Frank. The kid never realized what all the fuss with mommy and her friend was about. The Hollywood effect…

Frank once refused to be on a cover page of a glossy magazine. The magazine was trying to correspond with him in a manner he found rather amusing. 'Be prepared to allocate up to three uninterrupted hours of your time between 13:00 and 16:00 for a photo shoot. Send us your direct number so that the crew can contact you for the final details'. This is pretty much to the letter what was written in their email. Frank answered politely, but firmly, that he doesn't like to be on cover pages of glossy magazines and all appointments with him are made via the Public Relations services who schedule this kind of work. How was this possible, we were wondering later that day? Where did this unexpected tone of command rather than courteous request come from?

It dawned on us that, most probably, for the people who are 'private stars', this is a magnificent opportunity to stay in touch with their fans, to remind the world about their existence. The more they are seen and recognized by the public, the more is their professional value or monetary worth. Professional sports people often earn more from sponsorship contracts than from the award money for their performance. Their sponsorship value depends on their media visibility.

This popular magazine, with their knowledge of how individual celebrities appreciate their collaboration, corresponded with Frank on the same terms. It didn't even cross their minds that astronauts are civil servants, and that fame is a side effect (and not necessarily a desirable one) of them doing their work properly. Their work, like the work of top scientists, leading engineers or high-level military commanders, has a large bearing on world progress, and yet, their main professional audience, their 'stakeholders' (in correct language) or their 'fan club' (in popular language) – the people on whom their professional evolution depends – are not the general public. This audience is the people who train them and evaluate their readiness to do very complicated jobs.

I get a lot of 'sympathy' these days. People asking how is it to be alone for six months, while your husband is away in a potentially dangerous location. I'm a regular person, every now and then I enjoy a bit of attention, mostly when it comes from people who are part of my life already. Yes, I miss my husband, and yes, I would by far prefer to live uninterruptedly on the same planet with him. But I have to tell you honestly that temporary solitude, my way, is a five-star resort compared to the experiences of the families of nuclear submarine crews. Those crews go on secret missions where radio silence is a key condition for successful operations. Those people live for months in total isolation from the mainland, do their jobs and just have to assume that the world works fine. I don't think that I would have the strength and the spirit to go through something like that and retain my sanity. To my mind, those people are the real heroes. Are they in magazines? No. They are on secret missions. Are we their fans? No, because we don't know them.

So what made Frank change his mind? Why did he become a big proponent of writing a human story after all these years of firm rejection to talk about his personal life?

I believe this is because we have found an angle where 'human' is not necessarily a synonym for 'personal'. It might sound strange to begin with, but it's rather simple if you think about it. 'Human' is about human values and deeds. There is plenty to talk about in terms of traditions, tales and anecdotes. Some stories might be about the same events you, the reader, would have followed on the news, but told as seen through the eyes of the participants rather than a reporter. Other stories wouldn't have reached your attention and yet they are pleasant to share, they're fun and cheerful and not a secret. They don't intrude into anyone's privacy and give a much better feeling what life is really like on the space training scene and how very different it is to the artificial wrapping and superficial glamour that seems to be attributed to it by the external world.

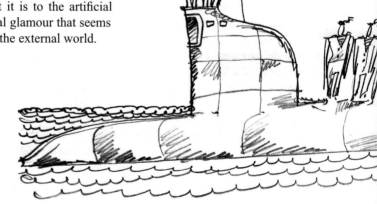

I have tried to discuss this notion of what is 'human', as opposed to 'personal', with some of the journalists who write about science. To my surprise, it wasn't easy to explain my point. Let me try again. Technical aspect: crew is launching to space. Human aspect: how do they themselves and their immediate families experience the launch (you will find out about this later in this book). Personal aspect: what lipstick was I wearing that day? Are you with me?

There is validity in sharing my experience of watching my husband blasting off into space, from the surface of the Earth, because this is a unique experience. There is no validity in my make-up choices because I don't happen to be a make-up expert. And if I was, this should be a discussion in the context of my professional engagement and not mixed into the experiences that I'm having in my personal life as Frank's wife.

There is validity in the discussion of human values of why Frank chose to become a good-will ambassador for UNICEF Belgium, which beliefs made him decide to work for this particular charity, but there's no validity in discussing, for example, what kind of car he drives. That would be a personal choice that does not have any relevance for giving a broader picture of space travel or Frank's human values.

Mechanics in garages have much more valid, interesting and valuable opinions about cars. Sadly, not many mechanics are often interviewed about their opinions. The most information we get about cars is published by the manufacturers who are trying to convince us that their product is the best. The same goes for what Frank and I, and our family members, do in our private time, where we shop for food, where we eat out, where we live, where do we spend our holidays, who are our friends, which newspapers we read and so on.

These days, the news seems to be mostly made up of scandal and drama. Every time Frank is asked for a 'human' story, he wonders if the newspapers are really looking to discuss his views and perspectives, and talk about global issues that are not necessarily related to the technicalities of his work, or are they looking to pull something out of context and make it look smashingly exciting. Unfortunately, this happens even in the headlines of technical interviews.

He once accepted an interview from a newspaper that declared itself as 'focusing on a human perspective' on the interview request form. He was looking forward to talking about the humanistic component of providing access to free education to underprivileged children in Africa. Instead he was bombarded with questions about the facts of his personal life. I think their first question was how he felt about the fact that it must be hard for his kids that he's always on the road and rarely sees them. I wonder what that journalist was expecting. I'm sure that every parent

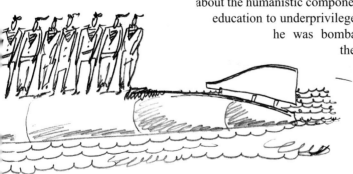

who works hard and travels a lot (or works part-time and does not travel but still thinks that they should be spending more time with their children) lives in a constant compromise. It does not depend on the specifics of the profession. It isn't interesting to discuss this because the answer is the same for everyone in a comparable situation. It doesn't add to the uniqueness of the human experience of being an astronaut. Astronauts are all human. They have the same struggles and the same doubts as anyone else. It's as simple as that!

And now, since the crew has trusted me to tell the human story of their flight (which effectively means the story of all our lives – them and us, the crew and their spouses), the tension and the discomfort of the interviews that might be misinterpreted is gone. Anything that should remain private won't be touched. If you're looking for news-breaking or scandalous revelations, you can stop reading now and put this book into paper recycling. Or maybe resell it with a comment 'Boring, but some nice pictures though'.

This is a nice, happy story about a bunch of people who were very lucky to share a most unusual experience. It has its ups and downs, as any human story does, but its mean value is always well above zero.

"The only thing I can say – you are amazing," I read in an email this morning from Frank, in response to my first few pages written yesterday.

I know he likes telling me things that make me feel good. But this time I'm also sure that he means it. He knows that now I will keep writing and, for the first time, together with his crew and their wives, he and I will openly tell one of his human, though not necessarily personal, stories out loud.

We all just hope that individual sentences and paragraphs are not randomly taken out of this story and turned into headlines that would sound absolutely wrong and misleading out of context, even though superficially dramatic and transiently exciting.

# 6 The ISS programme in a nutshell

## How is it organized?

The International Space Station (ISS) programme is an undertaking of five major partners: NASA for USA, Roskosmos for the Russian Federation, JAXA for Japan, the European Space Agency (ESA) for a number of participating European Member States and the Canadian Space Agency (CSA) for Canada. I can hardly believe that I'm writing this borderline bureaucratic introduction in the first line of the opening chapter of The Book that's meant to be a human perspective.

A great number of technical, scientific and properly positioned public relations materials are written and published about this programme. I could spare you formal language, and even the exciting, but by now traditional, facts about the ISS, but I actually think they are worth mentioning briefly, because it is unique and true. The 400 tonnes of metal, the size of a football field, the many billions of dollars of investment, the first and only global undertaking on this scale for peaceful purposes. I guess you know that much if you went as far as reading a human story about it. And if you don't, I would wholeheartedly encourage you to browse the websites of the mentioned agencies. The work they're doing in the various areas of space research is mind-boggling!

Nevertheless, I find it necessary to say in a few simple words how the ISS programme is organized. It's an important piece of information for telling this human story, drawing a landscape of what it takes to be a person inside this programme, whether as an astronaut or a family member. I'm sure there are also stories to be told through the eyes of the people who do other jobs in support of it. And I do hope that those will also come one day.

USA and Russia are the partners in the ISS programme who provide the biggest contributions. Accordingly, The Station consists of the Russian segment and the so-called United States Operational Segment (USOS). The USOS includes all the Western elements (provided by USA, Europe, Japan and Canada). It has docking ports that allow the docking of manned spacecraft and unmanned cargo vehicles.

Each partner has its own astronauts. Each partner trains all the space travelers who will work with the elements it has developed. In other words, all the work that has to be done on the Russian elements is taught to all the astronauts at Star City (near Moscow), the work on American elements is taught at NASA's Johnson Space Center in Houston and the work on Canadian elements is taught at the CSA in Montreal. Training on the Japanese elements is done at JAXA in Tsukuba, not far from Tokyo, and for all the European elements the training is done at the European Astronaut Centre near Cologne in Germany.

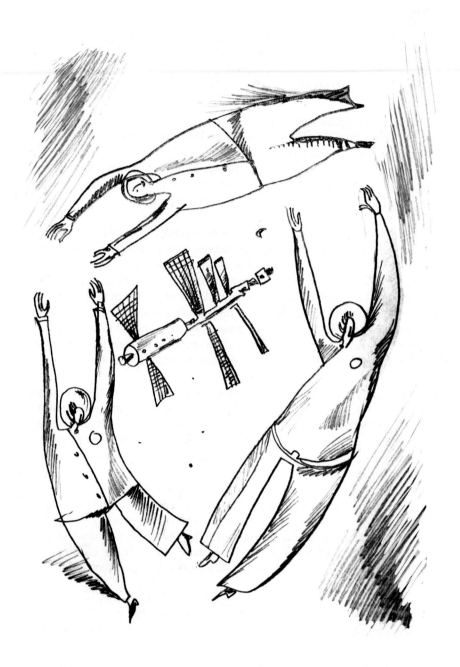

To be clear on terminology, Russians traditionally call their space travelers 'cosmonauts', and Western space travelers are called 'astronauts'. There are now 'taikonauts' from China and an 'angkasawan' from Malaysia, but there weren't any in the flight I'm writing about. So far, the Chinese have only flown on their own and they're not part of the ISS programme. Malaysia has had only one short flight to the ISS as clients of the Russian partner, as did Brazil and South Korea.

Depending on the intended flight assignment, the training to be received by the cosmonauts and astronauts is at the level of 'user', 'operator' or 'specialist'. Imagine you need to learn to operate a computer. You would be a 'user' if you knew how to use Word, Excel, Powerpoint and email. You would be an 'operator' if you knew how to download new programs and updates, and you would be a 'specialist' if you knew how to repair a broken laptop and reinstall the operating system.

INTERNATIONAL SPACE STATION

Depending on this level, the duration of training in each partner's location can vary, building up to three or four weeks at a time in Russia or USA, and six to eight weeks leading up to the launch at the location of the respective partner providing the launch. Every astronaut candidate undergoes a basic training of two years from the time they're selected, after which they're ready for flight training. When an astronaut gets a specific flight assignment, training can take up to another four years. In many cases, there are 'back-up' crews, meaning the astronauts who are trained in parallel so that if something happens to an assigned astronaut, they can be replaced by their back-up in order not to delay the launch. Often it's necessary to go through back-up training in order to receive an assignment as a prime crew member. Before his current flight assignment, Frank was back-up for Léopold Eyharts on the Columbus assembly mission. Columbus is the main European laboratory module on the ISS.

All the cosmonauts have Russia as their 'home base'. They live either in or nearby Star City. The majority of Western astronauts are stationed in Houston. Naturally, all of the US astronauts in training are in Houston, but so are some Canadian, Japanese and a few European astronauts. There is an unwritten 'gentlemen's understanding' that an astronaut should not spend more that 40% of their time away from home. This all works fine, until you turn out to be one of the two astronauts in the whole system who live in Holland. In this case, even the European training base near Cologne is not home. This is how we lived for over four years. Home in Holland. Training in Houston, Star City, Tsukuba, Montreal and Cologne. In general, training is a big chunk of business time, and almost 100% of your business time in the year leading up to the launch.

Currently there are two ways for people to fly to the ISS: with the Russian Soyuz or with the American Space Shuttle. Soyuz is a 'use once only' vehicle. There is a new one built every time people need to fly up and down. Together with Roman and Bob, Frank flew on Soyuz TMA-15. TMA is the current technical model, standing for Transport Modified Anthropometric. 15 is the sequential number of their vehicle in this series. During his first flight into space, Frank flew as flight engineer on TMA-1 with commander Sergei Zaliotin and second flight engineer Yuri Lonchakov. They came back in the Soyuz left by the previous mission, which was the last Soyuz TM model. Frank was trained for both models of Soyuz. That was a short mission. The Soyuz lifetime in space is six months and in that case their short flight had among its tasks to bring a replacement vehicle to The Station for the permanent crew.

Space Shuttles are reusable space planes. Currently there are three in operation: Discovery, Endeavour and Atlantis. Usually Shuttles fly and dock to the ISS and stay there for about two weeks.

Cargo for The Station is delivered by Space Shuttles, Russian Progress vehicles, and now the European Automated Transfer Vehicle (ATV) or the Japanese H-II Transfer Vehicle. While Progress is an 'old workhorse' that has been serving the space programme for a long time, the latter two have been specially built for the ISS by ESA and JAXA respectively.

There are three types of human mission that are currently flown to The Station. One is the Shuttle flight up and down, which is typically a 12–14 day mission. Often the Shuttle has a new component to bring up to The Station and the crew has a very intense schedule carrying out the task of fitting it. (A separate story should be written about the Shuttle flights to the Hubble Space Telescope. This is something that I find even more mind-boggling than the ISS programme, but this is another story!). Space Shuttles will be phased out in the next year or two. NASA will replace them with the new Orion vehicle which will be launched into space by the Ares launcher. It is designed both for flights to the ISS and for other human space destinations.[1]

Another mission type is a Soyuz 'taxi' flight. This means up and down with a Soyuz in a flight that typically lasts about eight days. Lately, only the private spaceflight participants have been making the taxi flights. 'Spaceflight participant' is the agreed term for people who are paying to fly. The term 'space tourist' was dropped by the space world after it became clear that the people who had flown not only demonstrated amazing skills in training and making the flight, but also developed and performed their own programmes for their flights and were big proponents of human space exploration.

The third mission type is an 'increment' flight. This is what Frank, Bob and Roman are doing at the moment. An exchange of the crew members on board The Station for long-term missions is called crew rotation. This replacement can be done both by Soyuz and by Space Shuttles. In case of the Shuttle, those crew members who are due to be left on the ISS are called by the charmingly amusing jargon term of a 'Shrek'. This is because they are the 'Shuttle Rotating Expedition Crew'. Or is it just because they are sweet and adorable? [2]

The Station was originally designed to have six people as permanent inhabitants and to receive other visiting vehicles and more people for short periods of time. It took a few years longer than planned to arrive at this capacity. Until Roman, Frank and Bob arrived at The Station on 29 May 2009, the permanent crew was only three people (on some occasions, only two) with occasional visits of three more from Soyuz flights, or six or seven with Shuttle crews. Together with Gennadi, Mike and Koichi, Frank, Bob and Roman became the first six-person crew.

To make The Station habitable for people, it has to provide the whole range of functions necessary for reasonable comfort and safe support of human life. The average temperature on The Station is 24°C. This means most people wear shorts and T-shirts. At the same time, you have to remember that in weightlessness, blood circulation (along with a wide range of other functions) is not assisted by gravity. This means that the heart pumps the blood but gravity does not pull it down into the feet, or into any other direction, no matter what position you are in. Because of this, some people find the temperature of 24°C to be still rather cold and wear long-sleeve shirts, long trousers and even thick socks and sweaters.

The amount of cargo that can be brought to The Station is limited by the capacity of the cargo vehicles. Cargo includes additional fuel for The Station. Fuel is necessary for maintaining its orbit. The Station loses its altitude and attitude very slowly as a result of various disturbances, such as atmospheric friction and Earth's gravity (altitude is the height of The Station above Earth's surface, and attitude is its orientation in relation to Earth). Every few months or so, a 'reboost manoeuvre' is performed by one of the attached vehicles to maintain the altitude and attitude. The Station flies in low Earth orbit about 400 kilometres above Earth. The forces in this orbit are such that while The Station is continuously drawn to Earth by gravity (in other words, it is continuously 'falling' to Earth, in 'freefall', effectively what is meant by 'weightlessness'), the speed at which it flies keeps it at the low Earth orbit altitude. To enter low Earth orbit at the required speed, you have to be launched to leave Earth at a speed of eight kilometres per second. If you are launched too slowly, you fall back to Earth; if you are launched too fast, you keep going away. For example, to fly to the Moon, the speed should be eleven kilometres per second (the 'escape' speed). So The Station flies at a speed about 28 000 kilometres per hour (about eight kilometres per second). If the speed in orbit was lower, The Station would fall to Earth, and if the speed was higher it would be flying further away.

Such continuous rotation around Earth is only possible outside the atmosphere, because the atmosphere causes a lot of resistance. This is why planes require engines to gain and maintain their speed to overcome air resistance. Their engines need fuel.

The atmosphere is formally considered to end at about 100 kilometres. In principle, this means 'space' starts there, above 100 kilometres. In reality, the atmosphere actually becomes less dense gradually, with its molecules (oxygen, nitrogen, whatever else is in the air) becoming more spread apart. Flights to about 100 kilometres will be offered by the mass 'space tourism' of the near future: a flight on a suborbital vehicle above the limits of Earth's atmosphere. This should give the passengers a few minutes of feeling weightless and an opportunity to see the curvature of the Earth. For low Earth orbit, 400 kilometres is the best compromise between atmospheric friction and fuel use. This is where there are still occasional molecules that affect the attitude to a very limited extent, but the fuel taken to correct the loss of orbital height is lower than if The Station was put into a higher altitude with even fewer occasional molecules (but that would require even more fuel to bring up the new modules and crews).

Another kind of cargo is water and oxygen. These basic elements are essential for survival. By now, the American segment has a system which fully regenerates almost all the processed liquids. Yes, in simple words, everything you sweat and pee is collected, recycled and you can drink it again. In the Russian segment, only condensate is recycled. It's used for 'technical water', meaning it is distilled water but not tested for microbial growth and is used, for example, to break down into oxygen for breathing or for cleaning the toilet. I think this is probably done for psychological reasons since, from a technical standpoint, the purification process is actually achieving good results. Turning recycled water into 'potable' or drinkable water is done by adding minerals and testing for microbial growth. Discarding liquids on board is almost impossible. No, you can't flush them down the toilet (as a matter of fact, you can't flush anything down the toilet, but we'll come to that later.) For this reason, it's not allowed to have foods on board like sausages in a tin. Because when you open a tin you need to do something with the liquid in which they are stored or pickled.

The general rule is that the only liquids you take along are those you will drink. Water is stored in reservoirs. If you want a drink, you can pour yourself water at room temperature or hot water (around 60°C). Imagine no cold drinks for six months! You pour your drink into a special plastic bag with a tube that allows you to suck it up and temporarily close it. If you want a fruit juice, tea or coffee, you take a plastic bag with the respective dry component inside. These are pre-packed plastic bags that contain a tea bag inside (it can be also with sugar – there is no way to add it afterwards), and the same for coffee. Arriving vehicles bring some fresh fruit that can survive a few days after the trip. In such cases, the crew gets to drink tea with lemon. This means biting into a slice of lemon and getting tea into your mouth afterward, or doing it the other way around. The choice is yours.

When it comes to fruit juices, nowadays there are no longer tubes with ready-made juice (you might have seen those in the past, they looked like huge toothpaste tubes). Today, rehydratable fruit-flavoured powder is sealed inside a plastic bag that you fill with water. Shake it and then leave it for about 10 minutes. This way it becomes a fruit juice. You should make up only as much as you are planning to drink. If you look at the footage of astronauts and cosmonauts playing with water drops, you will see that those drops are squeezed out of plastic bags. They have a perfectly spherical shape because of the lack of gravity (which causes it to take the shape familiar to us, much like a candle flame remains spherical and does not take the form expected on the ground).

You might also notice that after any demonstrations with floating water droplets they get 'eaten'. This is done not only for entertainment of the viewers, but also because this is a way to get rid of the floating water. Another solution would be to catch them in a cloth and leave it to dry. The water vapour will be caught in the water regeneration system and you'll probably drink it later after it has been through the full recycling.

A breathable atmosphere is maintained by the clever addition of various gases into the air. Those are brought up to the ISS as compressed gases.

The other cargo includes experimental equipment, spares for maintenance work on The Station and consumables for the crew. The calculations for consumables are made on the basis that on The Station one person uses about 3 litres of water (including drinking and all hygiene needs) and about 1.5–2 kg of food per day. Washing is done by wiping yourself with wet towels that are dried out for recycling afterwards. Brushing your teeth is also a special process. While squeezing paste onto the brush and the brushing motion itself is rather familiar, it's difficult to rinse your mouth with water because you can't really spit it out. First of all, it makes you think twice how much toothpaste you'd like to use – a great reminder that sometimes less is more. And then, after you take one mouthful of water, you prepare a folded tissue and spit it out in small chunks into the length of the tissue, so that everything gets soaked and does not escape. If it does, you have to go for a spit-bubble hunt. Everyone is obliged to clean immediately his or her own mess.

The bulk of the food for space is produced by the Americans and the Russians. The other partners have started developing their menus. Today it's possible to have some Japanese, Canadian and European food as well, but in limited amounts rather than a daily consumption. Food can be tinned or freeze-dried. Russian tins look like a regular tin you buy in a store. American tins are a soft foil, looking like a sealed silver envelope. To warm up the food, you place the tin into a special oven that will heat it up to 60°C. After that you open the tin with a regular tin opener in the Russian case, or with scissors for the American tins. The substance in the tin has to be of such consistency that there is no liquid left to avoid ballast and disposal problems, but it also can't be too dry: it shouldn't crumble when you eat it. Crumbling is an issue, not only because of the mess that flying pieces of food will create, but also because small particles might be breathed in or get into your eyes. In a worst case, you could choke on a floating crumb.

Another option for warm meals is to rehydrate the sublimated freeze-dried food with warm water (the same water you use for tea or coffee). Muesli and porridge are comparable with the regular earthly version, also rehydrated with water. Energy bars, dried fruit, nuts and honey are very similar to the ground ones but are packed into portion sizes – once open you don't need to reseal them. A general principle is that if you open something, you have to finish eating it.

Typically, daily menus are repeated every 10 days. On top of that, the crews get additional food of their own choice. They can ask for more items produced by the space food suppliers and they can also take some commercially available products, but in that case the same limitations apply. And of course, you shouldn't forget that there's no fridge or freezer for food storage, and you're not allowed to use glass for packaging. Food has to be suitable for storage at room temperature and have shelf life of over a year to get on board. Supplies for the upcoming crews are flown up before their arrival. The food and drink supply has to be placed on board three months ahead of the planned occupancy. This is important in case there are unexpected delays with the launches.

On average, personal consumables include one pair of socks per week, one clean T-shirt per week and one exercise T-shirt per week (clean T-shirts of the current week become an exercise T-shirt of the following week). There are a couple of long trousers and warm sweaters for the long-duration missions.

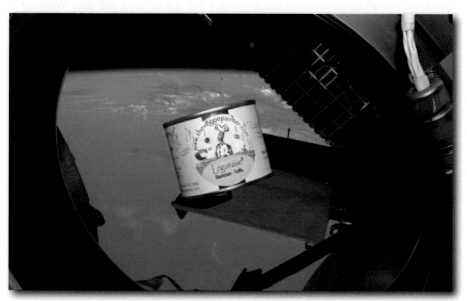

BELGIAN FOOD IN SPACE

Electricity is provided via the main solar arrays outside The Station. These can be rotated to face the Sun, which allows maximum exposure and accumulation of energy. External radiators allow excess heat to be lost and compensate for the extreme heat of the half-orbit in sunlight and the extreme cold of the half-orbit in the dark.

In anticipation of the first ever six-man crew, there were some uncertainties about the durations of the flights for various people. If you look at the masses and do the simple maths, you'll quickly see that one person 'costs' $(30 \times 3) + (30 \times 2) = 150$ kg of basic supplies, just food and water, per month.

# What does it mean to train for a long-duration flight to the ISS?

Astronauts and cosmonauts are trained in Houston, Star City, Tsukuba, Montreal and Cologne. The respective space agencies of the partners are kind-of 'Ministries of Space'. The manufacturers of The Station elements and the scientific facilities are major industrial companies. Through the years of training, people assigned to the spaceflights travel between all these locations.

## Star City, Moscow, Russia

Roskosmos, the Russian space agency, is located close to the centre of Moscow. Various companies involved in the production of spacecraft are located in Moscow, the suburbs of Moscow and several Russian towns. 'Star City', the Gagarin Cosmonaut Training Centre, is also on the outskirts of Moscow. Depending on the traffic, it can take anything from one to six hours to get from Moscow Sheremetievo international airport to Star City. If you are planning a visit in the winter, I would recommend giving yourself plenty of time for the ride to the airport. I know a couple of cases of people missing their flights in the crazy Moscow traffic by assuming that two to three hours should be enough.

In the late nineties, the Mir-Shuttle programme (American Space Shuttles flying and docking to the Russian Mir space station) became the precursor to the international collaboration and the International Space Station. During this programme, NASA built cottages for their crew members and staff residing in Star City. These few hundred square metres of Houston in the middle of a suburban Russian town are host to very important meetings and Fourth of July celebration barbeques, and give a place to live for those who are affiliated with the NASA part of the programme. But most importantly, they offer the environment of commonality and friendship. Where else would a bunch of extremely serious adults with very responsible jobs spend evenings cooking together, taking turns to wash the dishes, helping each other with the homework and simply spending the time in a big living room without feeling that you're imposing on somebody's privacy.

Unlike the trips to all the other Station partner locations, this six-cottage residence in Star City is the only one of its kind, where staying in touch with home, getting a call from The Station, having a work-out in the gym, cooking a meal, playing the guitar or watching a movie can happen within the space of several adjacent rooms. Where else would you find a communal home with a pool table, table tennis, a piano and an outdated jukebox all in one room; where all the inhabitants and visitors naturally and easily find something to do to enjoy their relaxed evenings?

There is always a NASA-assigned doctor following crew activities in Star City. Luckily for all the residents, most of these doctors happen to be good cooks. So are many astronauts! I doubt that it's a part of their job description and yet it tends to become a hugely valuable part of giving the home touch to 'life on the road'. Here people naturally take turns in preparing dinners. On most evenings, everyone can join an informal meal. The first time I offered to cook I was simply terrified. It wasn't a very busy week; there were no more than 20 people around, of which there were about 10 astronauts. Imagine the feeling though, of thinking what if something goes wrong with the cooking? Would the entire establishment come to a standstill because everyone had eaten from the same source the night before?! Of course nothing happened, but I tell you, I had to call my mother to double-check how to cook chicken, just in case!

The most memorable and amusing party in which I participated was a Korean birthday party. The short flight of a Korean astronaut was coming up. The real names of the two Koreans in training were extremely complicated for Western ears. Soon they were going by the names Sonja[3] and Kosan. It was Sonja's birthday. On the table with extravagant amounts of unfamiliar food (a big bowl of cold spicy noodles with a few pieces of ice thrown on top to keep it very cold), there were two magnificent classic cakes with white cream and different coloured cream roses. I couldn't wait for dessert! Then at some point when the main course was coming to an end, Sonja and Kosan gave a meaningful look to each other. Before anyone knew what was happening, Kosan had picked up one of the cakes and thrust it into Sonja's face. It took several seconds for the thirty people in the room to realize that both were laughing their heads off, and Sonja was totally happy with what was going on.

Apparently in some Korean university tradition, a birthday party isn't complete if the birthday girl doesn't get a face full of cake. I can't say that I regret that we haven't picked up this marvellous tradition for the future celebrations, but it sure was a treat to be there when it happened.

I know they're all working extremely hard, living on the road, missing families (I was extremely and exceptionally lucky to be able to travel with Frank a lot of the time) and yet those occasional wonderful leisurely evenings reminded me of the good years of full-time student life. The difference was that we could all afford good-quality food. And our parents only dropped in when we made the effort to bring them over.

# Visiting the Kremlin and Red Square

All the crews launching on Soyuz get to visit Red Square and the Kremlin, and to lay flowers beside the memorial plaques of Yuri Gagarin, the first cosmonaut, and Sergei Korolev, the 'father' of Russian cosmonautics (Konstantin Tsiolkovsky being the 'grandfather', he came up with all the theories and Korolev put them into practice).

Apart from Lenin's mausoleum, which remains the controversial centrepiece for honouring his role in history, the part of the Kremlin wall behind it is the memorial for the key figures in Soviet and Russian history. (A major acknowledgement to the re-evaluation of the significance of Lenin's role was made as the honorary guards were moved away from the mausoleum entrance to 'The Eternal Flame' war memorial along the other side of the Kremlin wall.)

In Russian there is an expression which literally translates as 'to be buried in the Kremlin wall'. The wall, a border between the Kremlin and Red Square, is effectively the most prestigious cemetery in Russia. This is the final resting place for the country's ultimate heroes. I personally question what Stalin is still doing there, but unfortunately I'm sure there are people who would fight me on that. The plaques on the wall carry the names of the deceased, and there's a structure in front of each plaque for flowers. Traditionally, red carnations are brought for all the occasions on the Red Square. The only way you can access the Kremlin wall in this area is by taking a tour to Lenin's mausoleum. The way out is a strictly defined path that takes visitors along the Kremlin wall.

TRADITIONAL PRE-LAUNCH CREW VISIT TO THE RED SQUARE

From 2009, there are about four Soyuz launches per year. In the past there were two per year and there were more Shuttle flights. Now, with the Shuttles gradually being phased out, the crew rotation will be done by the Soyuz vehicles. The two Soyuz launches per year were more or less in April and October. The launch of Roman, Frank and Bob, being the second flight this year but the first of the extra two, was at the end of May. The crew visit to Red Square coincided for the first time with the preparation for the Victory Day celebrations.

When I first came to Europe, it took me by great surprise that the British, the French and the Americans argue which one of them won the war. The first time I heard this, I was not sure how to react. It sounded like a distasteful joke to me. You, my dear European or American reader, are probably now wondering what I'm talking about (you may have been taught that your country won). You, my dear Russian reader, are probably now sharing the puzzlement bordering on disgust. How can anyone possibly question the role of Russia in World War Two, which for Russians is actually called the 'Great Patriotic War'. More than once in the 17 years I have lived abroad have I discussed this with different people from various countries and various levels of education.

In my younger days, I was even less of a diplomat than I am now. The 'little clarifications', such as on which territory was the biggest part of the war fought (for those who are not sure, the answer is Russia), how many Russian people died in this war (for those who don't know, over 20 million), where was the turning point (if your history books omitted this fact, it was at the Battle of Moscow), help in understanding the bigger picture of what is considered in the West to be an achievement of the US, British and French alliance. I have encountered these stories so many times that it became clear to me that it wasn't a joke and people were sharing what they honestly believed. Too bad that no matter how hard everyone tries to be objective, history seems to remain a propaganda-driven lesson everywhere in the world.

36

With all the exams and the travel plans, the only possible day for the traditional crew visit to Red Square was 7 May. It wasn't clear until the day of the visit whether it would really work out because the centre of Moscow gets blocked to traffic on very special occasions. Bringing military machines into the city centre for the parade and running the rehearsals of the military marches was obviously one of these occasions

Moscow is still short of decent public toilets. We discovered one right near the Clock Tower ('Spasskaya Bashnya') of Red square. The entrance is a few metres to the left of it. The temperature inside was about ten degrees lower than outside. It was a real dungeon. I could not help thinking of what it was before it became a toilet. Too many stories are told these days about the secret cellars of the Kremlin, the tunnels connecting the Kremlin and the secret underground prisons of the KGB building only a kilometre or two away, and secret lines of the Moscow metro purpose-built to provide additional shelter for the government.

I don't know enough to talk about it in any detail, but my heart still pounds when I have to face this subject. My grandfather Zakhar was killed in 1937. He was Deputy Minister of Culture. Like millions of other innocent people who drew a bad lottery ticket, he had some horrific case of nonsense fabricated against him. The family was told that he was in exile. The truth was that he was shot. In the nineties, when the whole Stalin crime scene started coming to the surface, my mother Lida, together with her brother Abraham, found out from the formerly super-secret archives that he was accused of something like 'ruining the library network of the country'. People who were shot were declared to be 'enemies of the state', and were officially stated to be 'ten years in exile without correspondence or visitation rights'. Therefore his wife, my grandmother Emma, was made to leave Moscow as 'the wife of the enemy'. She literally lost her mind. She didn't remember that she had a daughter and when my mother last had a chance to visit, shortly before her death, confused her for an older cousin. My mother was about twelve years old.

Having being born the daughter of a high-level government official, Lida was now turned into an orphan at the age of four. Fortunately, she was adopted by Abraham who had just turned 18 when all this happened. This saved her from being put into an orphanage. At the same time, this outrageous and unbelievable nonsensical cruelty probably saved Abraham's life. He was a young man on the eve of the Battle of Moscow. All the grown-up male population were drawn into the army but, as a 'son of the enemy', Abraham was seen as 'not worthy' to fight for the capital of his motherland and was left in disgrace to do clerical work. The majority of the people who fought in this turning-point battle were killed.

As I was writing this chapter, it made me realize that my family history probably plays a role in my non-acceptance of interference into my family life from any people in official functions.

After the flowers are laid at the Kremlin wall, the crews get a personal tour inside the Kremlin. In my Moscow student days, I did some work as a tour guide. I have been to the Kremlin on enough occasions to know my way around and yet I have never seen it completely empty. By the time we drove from Star City (Moscow reopened to traffic and this took at least a couple of hours) and went through all the prescribed ceremonies and picture-taking, it was closing time for the Kremlin visits. We were allowed through the Clock Tower, which is a special honour as only government cars and employees can go through there. We were given a personal tour of this magnificent place. I will leave it to the historians and their volumes of work to take you into the wondrous world of Russian art and legends enclosed inside those walls. For us, this was a special occasion because it signified the imminence of the space launch. The space memorial inside the Kremlin is the tree planted by Yuri Gagarin after his return from space.

LEFT TO RIGHT: BOB THIRSK, ROMAN ROMANENKO, FRANK DE WINNE, CHRIS HADFIELD, ANDRE KUIPERS, DMITRY KONDRATIEV NEAR THE TREE PLANTED BY YURI GAGARIN IN KREMLIN

Even though it was planned that the families would come for this visit, it turned out that Julia couldn't join us. Being the best wife in the world, she had to prepare the spontaneous party that she was told she would be hosting the day after. A bunch of Roman's friends from his military pilot school decided to get together to celebrate his upcoming flight. This might sound easy if you've only read the couple of sentences above. But the truth is that he studied in Ukraine and all his friends were working in different places around the world, so getting them together was as big a task as organizing an international summit. But so they did. One flew in from Canada and another cancelled a trip to India. They all got together for this very special occasion, to celebrate one of their own taking one step further into the sky, the sky that brought them together at the Chernigov military pilot school and made them inseparable friends almost half a lifetime ago.

ROMAN IN SPACE IN HIS MILITARY SCHOOL T-SHIRT

## Useful numerical superstition

There are a lot of useful superstitions in Russia. Some of them keep you occupied, others give you infinite pleasure and present positive conclusions about life. A special place in the hearts of modern Russians, who are by their roots inclined to mysticism, is found for various esoteric teachings or harmless alternative remedies. There is a reason why Russians are big consumers of completely useless food supplements. But there is also a reason why Russians are still capable of successfully curing a variety of simple illnesses without using any chemical medicines. For example, you can ease a cold by breathing steam from a pot of freshly boiled potatoes. You can rinse your throat with warm water and a few drips of iodine. You can even make mustard pads on your heels to pull 'badness' out of your body. Trust me, if you follow these procedures, they simply work.

Numerology happens to be one of those subjects that 'simply works'. Unlike boiled potatoes, which might not be for everyone, according to the ultimate source of the 21st century wisdom – Wikipedia – numerology is any of several systems, traditions or beliefs in a mystical or esoteric relationship between numbers and physical objects or living things.

In Julia's world, numerology works and makes sense. She doesn't spend much time on it, but considers herself familiar with the very basics and just calculates out of habit the ultimate number of everything that comes her way. Only one-digit numbers can be interpreted. You add numbers up until they end as one number. An example is the launch date of our husbands.

$$27/05/2009 = 2 + 7 + 5 + 2 + 9 = 25 = 2 + 5 = 7$$

Seven happens to be the best number you can get. I'm a bit less harmonized with interpretation of the numbers, but in my world this also makes sense. It's for a reason that seven is the number of good luck in various cultures, irrespective of their religion. It's for a reason that a lot of new beginnings happen in human life in seven-year intervals. I chose seven rays in the left lower part of the Expedition 21 mission logo for the same reason. Julia knew that everything is fine because the launch date was right.

As the theory goes, you can look into your own birthday and the birthdays of your parents to find numbers that influence your life. The digits in the birthday itself, as well as the total sum, should give meaningful information to those who are willing to see it. What can I say? So far it worked!

LEFT TO RIGHT: ANDRE KUIPERS, FRANK AND LENA DE WINNE, SERGEI ZALETIN IN THE STAR CITY, FEBRUARY 2009

# Big dogs

I love big dogs. There was a Doberman called Lada in my family when I was born. She thought I was her oldest puppy. She protected me from everything and everyone, but for the rest of the time completely ignored me. She was the older one of the pack and this gave her the right to be superior to me. As the story goes, at the age of two, I swapped sugar pieces for bones with her. It made us both happy but nearly gave my mother a heart-attack as she saw me walking into the kitchen, munching on a bone (note, it was a real bone, not modern dog food).

When I was four or five years old, I remember it was my idea of a good time to be rolled around in the snow by a dog bigger than myself. I was getting taller than Lada, so I chased other big dogs in hope that they would play with me. It never crossed my mind that they could do anything else. Big dogs were a symbol of friendship, love and happiness. I am definitely a big dog person. To date, I'm inclined to believe that if in your sleep you dream about friendly dogs, it's for good luck and friendship. What about cats? I'm not against them, given that one gave birth in my pram one night when it was left outside in good weather. My poor Mom…

There are a lot of stray dogs in Moscow. There's been a long public debate on what to do with them. Animal rights activists claim that poor dogs need to be protected, since it was not their fault that they were born. I have never been an activist of anything (apart from the human rights of the wives of the astronauts, because of how I was treated at Frank's launch) but I always had a weak spot for dogs. Big dogs were friends. Mixed-breed dogs were intelligent. My father told me that if I ever had a doubt about whether a stray dog was a threat, I should pretend to pick up a stone and throw it. Apparently all stray dogs recognise this gesture as a promise of a stone thrown at them, and they should run away. He was once faced with a pack of stray dogs and this was how he managed to escape, by becoming aggressive in return, threatening them with his imaginary stones and overpowering body language.

I love big dogs. I remember I once bought some food in the street and gave it to a big scruffy stray dog. He was looking at me and his eyes were too sad. When I stretched my hand out to give him the food, he swung away. It was obvious that he was expecting to be hit. I squatted to get down to his eye level, so that he could see that I wasn't threatening him. He started eating, and I walked on. Occasionally, one dog would follow me for a while. With these dogs, I felt like we'd known each other from another time and finally met (the last time this happened was literally one month ago, during my brief visit to Moscow while Frank was in space). But what could I do? I learned eventually to keep going instead of spending time talking to them. I couldn't take care of them. I didn't want to mislead them.

I refused to form an opinion on what to do with stray dogs. The city should be clean, but they were living now and it wasn't their fault that they had been born. There is a 'humane' programme to catch, neuter and release these dogs, so that eventually they die out from natural causes. Of course, this hypocritical approach doesn't work. Even if you assume that all the money allocated to the implementation of this activity is spent on it, you just need to miss a few of them, and the population will be maintained. According to some analysis, one female together with her immediate offspring can produce about fifty new dogs just in one year. It's very profitable to maintain a funded neutering programme, because it will never end.

Even though I grew up in the centre of Moscow and seeing many stray dogs was not strange to me, I'm not so comfortable anymore with this idea after 16 years in Holland, where animals not only belong to their owners, but also are not supposed to be let off the leash in public places.

One night in February, I was walking back alone from Cottage 3 at Star City back to the 'profi' (profilactorium) building. Frank left straight after dinner because he still wanted to study for the next day, and I stayed in Shep's bar listening to Chris Hadfield playing the guitar and Cady Coleman playing the flute.

The bar was named after its 'builder', Bill Shepherd, the American commander of the first Expedition to The Station. Bill had pretty much built the bar with his own hands during his years of training at Star City. Chris and Cady's diverse repertoire had tunes for every occasion. Their coincidental presence in one location (Chris was in training as a back-up for Bob, Cady training for her flight a year later) added a dimension to those special evenings that otherwise stayed unexpressed without their music.

This February was in a real Russian winter with a lot of snow. The only areas not snowed under were those specially cleaned or those where the hot water pipes were too close to the surface of the ground and melted the snow above them. As I walked about 50 metres in the direction of the 'profi', reaching a point exactly between the two cottages, far from any door, I saw a pack of about ten dogs emerge from a snow-free strip in front of the next cottage. It was close to midnight and they'd probably settled for the night on warm ground.

As far as they were concerned, I was trespassing. As far as I was concerned, I was on the territory of a space organization with a secure entrance, which supposedly provided safe accommodation for cosmonauts and astronauts up to two weeks before their launch. Being bitten by a stray dog shouldn't be in the schedule, you would have thought. Wrong.

Spaceflight participant Charles Simonyi was bitten in bright daylight as he entered the food shop about five weeks before his second flight. He didn't even see the dog approaching him! If you could call anything like that lucky, it happened early enough luckily for him to get the necessary treatment and not to miss his flight because of a dog bite. What this boils down to is that you can lose a flight because there are stray dogs living inside Star City, and it's impossible to get rid of them. Even if you manage somehow to get rid of those few packs that live there now, word would soon get around to other wild dogs and they would sense that the territory is empty and start claiming it. It seems impossible to end this vicious circle until a general approach to this problem is defined.

I froze. I knew that the only way to make things worse with aggressively-minded dogs is to move. I only had my coat and a tiny mobile phone in the pocket. I tried to slowly imitate picking up a stone. It didn't have the desired effect. If anything, it made them step in my direction. I was part erratically thinking what to do, part observing them. I couldn't help but notice though that their young puppies were adorable. While the big dogs definitely saw their encounter with me as a serious activity, the puppies with their fluffy paws and those huge ears were wagging their tails in eager anticipation of some fun. And yet, for the first time in my life, I was afraid of dogs. I tried to make small steps backwards without losing eye contact, making sure that it didn't look as if I was moving. I was afraid to pull out the phone, and I wasn't sure who to call anyway. By the time anyone could get here, the situation could have resolved itself in any number of ways. I'm not sure how long this lasted, but it didn't feel a short time for me. It was probably a minute or two. Extremely fortunately for me, two American guys came out of Cottage 3 to go to another cottage where they were staying. As they approached, they were shouting and waving their coats, making themselves appear big and superior. The dogs yapped a couple of times and reluctantly ran away.

In the meantime, the debate continues and will probably never end. Old grandmothers feed dogs in their yards. If anyone tells them to stop doing it, they get upset and have a go at the 'offender' with a fashionable folk wisdom that 'dogs do not harm good people'. I still love big dogs. But it's clear to me now that urban dwelling is for people.

People should keep animals as pets only in a responsible manner. Other animals don't belong there. I'm sure it took time to bring Europe into order. As harsh as it might sound, it's time to do something about it.

## And this also happened

André Kuipers, an ESA astronaut from The Netherlands, was training with Frank as his back-up. Normally the process is organized in such a way that a back-up astronaut on the current flight becomes the prime crew member for the next flight. This is what happened in Frank's case. He first trained as a back-up to Léopold Eyharts for the previous European mission, and then became a European crew member for this flight.

Somewhat unexpectedly, at the end of April it looked as if political circumstances were going to change and André wouldn't have a chance to fly anymore – yes, just because of the politics. If you look closer at this situation, you realize that it means much more than just 'not getting to fly'. In the same position as us, with his home in Holland, André was on the road for every single training session he took. After several years of spending most of his working time travelling, away from his wife and kids, it could turn out that all this personal and family sacrifice would have been a waste. For some months, the situation looked rather grim for André. Fortunately it was rectified and he was officially assigned in August for a flight in 2011. But those months were seriously tough.

FRANK TOOK A PICTURE WITH HIM TO SPACE OF THE CHURCH IN STAR CITY

Chris's guitar and Cady's flute were there throughout the years of training and almost right up to the time of the departure of 'our' crew to Baikonur in May. Chris and Cady were playing their music in the evening of that uneasy day in April when the news came through for André. We all thought at that time that there would be no solution for him. There was not much anyone could do, apart from offering good company. It was close to midnight but we wanted to stay together for a little longer. We went outside. As it often happens in spring, when the night is clear, it gets very cold. This night was no exception. But it didn't seem to matter. This time, slow and somewhat melancholy tunes from the flute and guitar were pouring out into the cold starry night, rising above the cupola of the unfinished church and shimmering in the middle of the field. A couple of church workers came out of their little hut and joined us in silent contemplation. Frank had exams first thing the next morning and was already sleeping to be well prepared. I danced alone to the music of this crisp night.

## 'The Planets'

Cady's husband Josh Simpson is a glass artist. What a beautiful match – an artist and an astronaut!

Apart from magnificent handcrafted pieces of tableware and vases, Josh makes a variety of different sized spheres that are called 'The Planets'. Through the transparent surface, you can see magic terrains of alien but friendly worlds, playing with all colours of the Universe, reminding us of the beauty of the Earth and yet leaving a prying eye no illusion about its extraterrestrial nature. Josh and Cady have a wonderful game that they have been playing for years. They occasionally hide the smallest 'planets' (which are just over an inch in diameter) in various unexpected locations all over the world. If someone finds them, good for them! I had this most exceptional privilege to accompany Cady in hiding a little planet somewhere in Star City. Who knows, one day you might be the lucky one to find a wonderful surprise. But I'm not going to help you. I know how to keep a secret!

## Houston, USA

"Beautiful city," said one of our friends a few years back, with a lot of sympathy in his voice, in response to my story that I'd been to Houston for the first time. He gasped for air when I said that it was in July. As I stepped out of the airport building, the wave of damp heat covered me from head to toes. "This can't be right," was my first thought, because I'd grown up with an occasional unreasonable frost, but not with such a heat. A hundred percent humidity is dangerous, not only because it's accompanied by this unbearable heat, but also because in this condition you can't sweat: it's impossible to secrete liquid into an atmosphere fully saturated by liquid already. To my great surprise, the world managed to keep moving at a normal pace even though you would have thought that these temperatures affect all living beings and would drive them into some kind of overheated hibernation. Truly there are no limits to the human ability to adapt. A couple of days later I understood the origin of the American national game of 'find the closest parking spot to the entrance'. It was a matter of arriving at the next door completely soaked in your own sweat or at least looking half-human.

The premises of the National Aeronautics and Space Administration (NASA) are situated in various states, including its Headquarters in Washington, Kennedy Space Center (KSC) at Cape Canaveral, Florida (where the Shuttle launches take place) and Johnson Space Center in Clear Lake, Texas, which is the main training location for all the American elements of The Station, and this is the ultimate destination for space training trips to the USA.

Visitor centres affiliated with the major NASA sites host a wealth of knowledge about the history and vision of space, offer virtual space trips, educational experiences, opportunities to buy various space memorabilia and, in the case of JSC, an opportunity to take a guided tour into the heart of the hearts where it's all happening: the visitor gallery above the training hall with plenty of astronauts in training for their spaceflights.

In the last trip to Houston before the launch, I started receiving all the Expedition 21 crew personal souvenirs that we'd decided to produce privately. It's traditional, for example, for the Shuttle crews to make rather expensive jewelry for their immediate families and close friends. Astronauts privately spend quite a lot of money to create memorabilia for their flights. After getting a lot of advice on what others have done, we decided to make a very small lapel pin of the Expedition 21 patch and give it out as a special gift from the crew to all the people who worked with them. I'm really happy that the plan is working and there are lapel pins out there now that you can wear as a small symbol of having been a part of this mission.

Apart from the pins, which were the key item, I also ordered a few tie-clips, earrings and charms in the shape of the Expedition 21 logo. With great astonishment, I realized how much attention space memorabilia gets in the collector communities. Nicole Stott was wearing her Expedition 21 earrings in the first crew conference in the beginning of March. The next day a posting appeared discussing this on one of the space collector websites! The charms were mainly used by us as decorative hangers for mobile phones. But the most important charm of all turned out to be the one worn by Chris, Bob and Brenda Thirsk's dog.

## Chris, the Dog, the First Bearer of the Charm
*(Special contribution written by Brenda Biasutti-Thirsk)*

Several days before the family's departure for Russia, Chris, our miniature schnauzer, did two things that he is very good at. He had me worried sick and then laughing out loud, in that order. I had been out running errands and came home to a chaotic household, composed of my three children and a hysterical dog. As soon as I walked in, the kids reported that Chris was hyper and I knew right away that something was wrong. My beloved dog is not hyper. He rests and naps and sleeps and only fetches twice in a row before confusion and exhaustion set in. I love him enormously however, and accept his lifestyle and understand his tendencies towards energy conservation. When he greets me at the door, he's usually sporting a crushed beard and a crooked smile that give him away. But that night he bounced around us, with energy that seemed to be borrowed from elsewhere. I was perplexed but decided to wait before taking any serious action. I contemplated the dilemma I would be faced with if Chris was very ill. I could not leave him, but also could not postpone the launch, nor did I really want to.

My mind quickly wandered to other matters that were beckoning the night before our big trip. I walked into my bedroom and my eyes were drawn to some scraps of paper on the floor. On the carpet were two wrappers from some special chocolate that I was planning to take to Bob. I had bought two large bars of organic chocolate that were supposed to contain love messages inside.

I had left them in a large shopping bag with the intention of packing them that night. Instead, what I saw were empty wrappers imprinted with Chris's teeth marks! I could not believe that the dog would eat so much chocolate without leaving a crumb. I feared the worst, knowing well that chocolate could be lethal to dogs.

I went running into the living room, informing everyone of what I had just discovered. And then it happened. Chris began to gag convulsively and proceeded to regurgitate two whole bars of organic chocolate.

By the time I was done cleaning the mess, Chris was stretched out on the sofa in his usual and more subdued pose. My children later came out of hiding and we all laughed at the drama that had distracted us from the tedious task of preparing for a long trip. Chris had again succeeded in reminding us of the importance of laughter to dampen the everyday stresses. He is the emotional thermostat that smoothes the inevitable wrinkles in my life.

Before the launch, Chris acquired an honorary degree that he wears with pride. Lena had taken the charm from her cell phone and put it on his collar the night she met him. The charm bears the labour of her design for the crew patch of Expedition 21. That night, Chris was assigned to spread a significant theme that was embraced by the crew of Roman, Frank and Bob, and manifested through the relationships that were solidified over the course of the mission's preparation and evolution. Lena took the lead in ensuring that the symbols of collaboration and connectedness presided amidst the technical objectives of the mission. Chris carries on him a charm that embodies the spirit of relationship, love and humanity. He does so faithfully and unconditionally.

# Big astronaut family

No matter how hard you try to anticipate and prepare for forthcoming things in life, it always turns out that the reality is rather different from the results of your mental exercises, both in terms of your feelings and because of the novel circumstances you can't foresee.

One of the things that I never expected a few years before the flight was the number of people who have nothing to do with Frank but who would want to 'touch a celebrity'.

A peacefully mad-looking man in the centre of Brussels kept insisting on discussing the presence of a UFO in the vicinity of Earth (it was clear to him that Frank had seen them but was not allowed to disclose this highly secret information). The most frequent occurrence of 'touching a celebrity' happens when we pass by the town in Belgium where Frank used to live. Questions like "My grandson went into the same military school as you but only five years later, he had a moustache in those days and was wearing glasses, do you remember him? He says you were best friends then."

Frank always agrees that he remembers blondes, brunettes, bald or otherwise, with or without moustaches, with or without glasses. He's a nice person. So he pretends to know what they are talking about. But the one thing that came as the biggest surprise, and became progressively more burdening in the year leading up to the flight, was the big private party. What I mean by this is that if a friend is hosting a birthday barbeque for example, and invites twenty or thirty people, no matter how easygoing this private evening or weekend event is, it unavoidably turns into a PR event for Frank. Most people present sooner or later will start asking "How is it up there?" and the bravest ones take a deep breath and ask about going to the toilet. Most want to take pictures and do all those usual things that are appropriate and expected when an astronaut makes an officially scheduled public appearance as a part of the job.

At this point, a party that was meant to be a private dinner for us becomes an activity similar to when I join Frank in his official appearances. He answers the questions that keep being asked over and over again, and I have to wait for the collective adrenaline rush to settle down. My experience shows that it does not really happen naturally. So at some point we make an effort to leave. That way the party turns back into being a party and not a celebrity PR event. In the year before the launch, we had to recognise that private evenings in this format became a job. What could we do? The solution is simple. Back in Europe, we now only organize or join the private parties where everyone knows each other and have no Hollywood effect on each other.

From this perspective, Houston seems to be an easier place for being a family with an astronaut in it, because there are plenty of those around. This doesn't mean that you are all friends, or even are in touch. But it does mean that you get an opportunity to belong to a rather large group of people where the normality of your private being is a common denominator and therefore is not challenged. You're surrounded by a big group of people who are having the same firsthand experiences as yourself: the same dreams, the same fears, the same aspirations and the same endless years of waiting. First waiting to be assigned to a flight, then waiting to do years of training, then waiting for the launch, then waiting for the flight to be safely over and then waiting to rebuild a normal life. Normal only from the point of view of those who are living it. Because if you peel the thin layer of glamour that comes with the position, the challenges, limitations and compromises that the families of astronauts face are rather substantial.

In Houston, you can dare to speak openly about your problems and feelings and expect reasonably that you'll be heard and understood. Only when someone has been through what you're going through, can the real connection, comprehension and supporting bond develop in your relationships.

When I was discussing this subject with one of the astronauts who lives and works in Houston, he immediately agreed and gave me an example of his son who had been on holiday in some remote location. As soon as the bunch of kids he was with found out that he was a son of an astronaut, the question came up from the 12-year olds, "What's it like to be a son of an astronaut?" The answer was magic in its simplicity: "I don't understand what you mean. All my friends' fathers or mothers are astronauts. It's normal. What's the big deal?"

Have you ever thought that offering help is much easier than asking for help? Giving help, especially when it doesn't take you far out of your way, is an easy way to remind yourself that you're a good person. You get a nice cuddly feeling. Asking for help, on the other hand, is exposing your weaknesses and vulnerabilities. If you're honest with yourself, I'm sure you would agree that there's only a very limited number of people on this planet that you can trust to know your weaknesses and yet feel safe and protected in their company. This must be why the American culture is full of shrinks. It's unhygienic to be weak. You pay to deal with your doubts and hesitations in a professional environment and walk out of the shrink's office on a weekly basis, all polished and ready to take up new challenges as a winner.

In Houston, if as a wife of an astronaut, you have moments of being deflated and need support you can either go to a very competent shrink who understands the specifics of your problem, or go to those who are in the same situation as you, and who know exactly what you're going through, and whose company would make a difference to both of you.

One of those people in my life is Brenda. We didn't get to spend a lot of time together before the launch, because she was in Houston and I was in Star City. We really started talking only after we both got home in June.

Life on the road, like many things in life, has advantages and disadvantages. One of the greatest advantages is that you live in an international community and very quickly learn that different doesn't mean wrong. Differences come up in various forms of insignificant daily routine, but if they remain misunderstood, they can bring discomfort to people. Starting with the standard greeting, "Hello, how are you?" and conventional response, "I'm fine, thank you, how are you?" and going further into accepted body language, 'personal space', good manners, you name it. Russian men maintain their courteous ideas about opening doors and helping with heavy bags. I love this, and I see it as a sign of friendliness rather than disrespect of my persona.

Subtle cultural differences are present when it comes to what constitutes acceptable greetings between the people who know each other well. In the West, everybody shakes hands. In Russia, women don't shake hands. This is a manly gesture, coming from the days of the mediaeval tournaments when knights would remove their gloves before battle and shake hands to show that they were not hiding any unfair weapons. You can shake hands with a Russian woman at the most when you meet for the first time as a part of introducing yourself. It took me a while in the early nineties to figure out that people who kept offering a handshake had not forgotten that we had already met, but were being polite in saying hello. In America, the done thing for greeting people you're close to is a distant hug. Or they just heartily pat you on the back, unlike some British people who kiss the air next to your face with a 'muahh' sound. This was probably the reason why I started signing all my informal correspondence with 'Big hug, Lena'.

I type fast but not very accurately. 'I' and 'U' are next to each other on the keyboard, as well as the other adjacent letters, and they get in each other's way in my text. Too many times I've written 'bug hug' and felt compelled to correct it, until one day I thought that it was special and was happening for a reason.

"Hey Brenda," I added at the end of my note to her where 'Bug hug' came out naturally despite my concentration on correct typing, "If you don't mind, I will not try to correct this typo anymore. You know that I meant 'big'. But it keeps coming out this way."

Brenda loved this idea. We now send thousands of bugs with love and hugs and support to each other. No matter how far apart we are, we keep each other hugged, especially when one of us feels completely bugged. She gave a name to our life during the flight and to her folder where she files our correspondence: 'Operation Bug Hug'.

## Montreal, Canada

With only very little to tell about our travel there, Montreal nevertheless is a nice city to visit. Ironically, I found there, what for my taste was, the best sushi restaurant in the world. And we acquired a traffic fine. I still remember it because it was totally unfair. We made a left turn following a police car that was driving in front of us, but, true enough we did not pay attention to the road signs. That very police car that we followed stopped immediately after the turn and gave us a fine for crossing a double line (hidden under the snow). Apparently the rules applied to us but not to him. Oh well. If only all the problems could be paid off with one hundred Canadian dollars, the world might become a better place. I realize that it sounds almost too trivial to share this on the pages of The Book about a space mission, but it shows over and over again that human life remains down to Earth, no matter what the focus of your professional commitment.

The Canadian Space Agency is called, amusingly symmetrically to my Russian eye, the CSA–ASC, because everything official in Canada has to be in two languages, English and French. The organization is small and efficient. One classroom and one simulator room are conveniently located next to each other. This is where the training classes take place for operating the Canadian Robotic Arm (the Canadian contribution to the ISS programme). State-of-the-art training technology perfectly represents robotic operations on board The Station. Robotic operations are also trained for in Houston.

On one occasion, Frank brought me a plastic wrench from his class at the CSA.

"This is cute," I said, easily owning up to my female reactions.

"It was printed on a 3D printer," said Frank.

"It was what?" in my world the word 'printing' has a very particular unambiguous meaning.

"This would be a perfect technology for us to produce spares on board," Frank continued, still saying things that remained as clear misfits in my trivial 3D world.

"Look, this is so great in its magic simplicity!" he said with a happy sparkle in his eyes, as he does when he finds something new and genuinely progressive. "The printing is made of polymers. Two colours, yellow and blue, made up layer by layer. Afterwards the resulting block is put in a solvent and the blue colour gets fully dissolved, and the yellow part remains as the item. As in this wrench, it can have moving parts."

Somehow I was more impressed with this plastic wrench than with the whole space programme. Aren't we all creatures of our habits?

Frank has been for training to the Canadian Space Agency in Montreal a couple of times for a couple of weeks to study the operation of the Canadian Robotic Arm. With his background as a military test pilot, opportunities to operate tricky machines that

FRANK WORKING WITH CANADIAN ROBOTIC ARM IN SPACE

are off the ground are his idea of good time. Just to clarify what it means to be a test pilot. An aircraft has a performance 'envelope', meaning the limits of its design in terms of airspeed, loads, altitude or other capabilities such as maneuverability. Under normal circumstances, the flight is performed more or less in the middle of this envelope. Testing an aircraft includes operating it also on the margins of its performance capability. Detailed procedures are developed by test engineers and implemented in flight by test pilots. For example, one task for testing an aircraft might include shutting down the engine in flight and then trying to restart it again. If the engine can't be restarted, the test pilot will have to make an unpowered landing like in a glider.

While 'pushing the envelope' is prescribed in the flight plan in order to validate its design parameters, it can't be ignored that an aircraft might exhibit some unpredicted behaviour or suffer some kind of failure. Then the test pilot would be expected not only to report the failure or unruly behaviour, but also to give accurate, structured and quantifiable data describing what happened. When I hear things like that, my idea of good time is to hide under a big pillow and to never travel by air again!

There is equipment on The Station for maintaining proficiency in the various skills learned during training on the ground. The robotic simulator is one of them. Robotic operations in space, along with the spacewalks, are at the top of the list of the most complicated things to do. In anticipation of the maiden flight and arrival of the Japanese HTV cargo vehicle at The Station, a lot of onboard refresher training was prepared to keep the crew in top shape for this immensely important event.

And then the robotic simulator on The Station broke down. With his background in flight testing and reporting on in-flight failures, Frank was able to give feedback on what was going on and, under the guidance of the ground engineers, take further steps in generating more data. As a result of this collaborative effort, a software patch was developed on the ground that Frank then installed to recover the simulator. Yet again, his experience of being a test pilot and an engineer became priceless in generating quantifiable data in a team effort for searching for a solution.

However, unlike the breakdown and repair of the failed toilet, about which you will read in other chapters of this book, the repair of the robotic simulator didn't get any media attention or coverage – like a few other equally complex and critically important breakdowns and repairs that took up a lot of time and real work between the ground and The Station crew. It made me think again about the superficiality of the 'fame' glow surrounding the astronauts and their work, the 'Hollywood' effect of connecting to a celebrity.

Technically, both the toilet and the robotic simulator breakdowns were unforeseen malfunctions. Operationally, in both cases, it was as a result of the interaction between a crew member and ground engineers that some very special work was developed and implemented in real-time in order to recover critically important equipment, the failure of which could have brought detrimental results. The only difference is that, in one case, some repair work was done to keep the robotic simulator in order and in another case it was done on a toilet (except that the repair of the toilet was much simpler and less critical). Everybody is always interested in the evolution of the toilet. No one, apart from the specialists working on it of course, even thought of asking how the recovery of the robotic simulator was going.

# Kyoto, Tokyo, Tsukuba, Japan

I didn't know what to expect from Japan, but whatever I experienced during this two-week trip was certainly very different to anything I could have expected. After literally a few minutes in the Narita International Tokyo airport, the phrase 'Germany of the East' came to mind (the same as St Petersburg is the 'Venice of the North'). The level of orderliness and prescribed impeccable correctness was off the scale by any European standards. In Europe, it is considered that Germans are the most organized and orderly of people. Their disadvantage is that they get corrupted by not-so-well organized neighbours. Japan, being an island, on the other hand, as it looked to me had managed to retain its pristine and authentic uncorrupted straightness.

Passport control. Pretty girl in serious-but-pretty-nevertheless uniform. Someone told me before the trip not to stare into the eyes of the officials. This is counter to all the other borders I've ever crossed before. The pretty girl only briefly glanced at me and shyly looked away. I glanced at her once and then started staring at everything else around. Click-click, stamp-stamp, "Welcome to Japan!"

Just a short stretch of water between Japan and Russia. And what a difference! As I'm writing, it came to my mind that I've never heard when crossing the Russian border in any direction, "Welcome to Russia" or "Have a safe trip," either as a former Russian nor now as a European. Customer-oriented service at its best. An ancient joke from the Soviet times comes to mind. A foreigner arrives in Russia, exits the building of the airport and almost falls into a hole in the ground that is not marked:

"You should put a rope with red flags around that, it's dangerous," he reports to a nearby policeman.
"Have you crossed the border of this country?" asked the policeman.
"Yes."
"Then didn't you see the big red flag at the entrance?"

Back in Japan, after the shyest passport control in the world, there was customs. We were on a two-week trip.

"What are you bringing with you into our country?" asked a very serious-looking man, with an expression that in European culture would have sounded like concern. I didn't think that I looked like someone who could cause concern in a customs officer, so I attributed the seriousness of his expression to my inability to read it.
"Clothes in this one and some books, shoes, toiletries in the other. Personal

belongings and computers in the hand-bags," I stated.

"I will check this one bag," he pointed at the bag that I'd identified as containing clothes only. His face expressed the commitment to protect his country from any illegal clothes.

"Of course, go ahead," I opened the bag.

It was full of clothes. The face of this dutiful officer expressed what in my body language would have indicated the deepest level of disappointment.

"I promised to check only one bag, so I check only one bag. Welcome to Japan."

We had an opportunity to take a private weekend and visit Kyoto. We had to buy tickets for the 'shinkansen' bullet-train to Kyoto. The unbelievable disadvantage is that it's almost impossible to pay anywhere with credit cards. The super advantage is that the trolleys that you use to move your luggage have such a construction that you can comfortably take them on the moving staircases to the underground train level. To enter trains, people waited in a straight line that they formed as they arrived. Not a single person (apart from us) tried to get closer to the platform faster than the natural progression of that queue. By the way the locals were looking at us, we quickly realized that we were doing something completely out of the ordinary.

CONTROL PANEL OF A JAPANESE TOILET

The conductors on the train entered the compartments with a bow and an incomprehensibly long and humble greeting. They checked the tickets and exited with the same bow and the same greeting. The toilets on the train (as well as in many other tourist locations) come in two types: local and European. (Look at me, after all the jokes I make about people being interested in space toilets, I find myself thinking the Japanese toilets to be one of the most interesting cultural experiences of the whole trip.) The local toilet is what we in Europe would call the 'French toilet' – pretty much a hole in the ground. However, the European toilet is something that appears mechanically to look like a regular toilet in our terms, but it has honestly more controls than my car! As a matter of fact, I have a confession to make. The first photograph I took when I landed in Japan was a photograph of the control panel of the toilet in the airport. Not only did it have a heated seat, but also a button to create the artificial sound of falling water!

Toilets and manga comic strips in Japan

made me think about the Freudian hyper-compensation. In the past, their toilets weren't really comfortable to use. So they developed high-tech equipment that gives you all sorts of physical comfort and relaxation as an integral part of a very important physiological process. Oriental faces have a particular shape, especially the eyes. All their cartoon characters have the eyes of a Disney Bambi deer. Japanese women are mostly tiny and on average significantly smaller and less-curvy than non-oriental women. All their comic-strip adult females have disproportionate waist curves and rather impressive breasts.

I love Japanese food. I might have been Japanese in one of my previous incarnations. I didn't understand a word and yet I felt totally happy. After this trip, I had so much energy that it felt as if I had visited home and recharged my batteries. Frank, however, likes meat and other regular 'male' food. After two and a half days on rice and fish, which made me ecstatically happy, we had to look for 'real' food. The solution stared into our faces in the form of two painfully familiar golden arches. There were only the two of us and what seemed like every teenager in Kyoto eating there. With the only difference that we were trying to have a conversation and they all had their ears plugged and hands occupied in text-messaging or game-playing.

Japanese exhibit all the food they serve in the windows of their restaurants. The resemblance is amazing. You get exactly what you see on a plastic plate. This must be a serious line of business: to produce menus in plastic. I saw an absolutely marvelous ice-cream café with what looked like at least one hundred different combinations of ice cream, fruit and delicious-looking sauces. I wanted to take a picture. But it was absolutely not allowed. The owners stood there and watched me to make sure I went away and didn't try to take a photo in secret. By the time I found a shop with hundreds of delicious plastic cakes, I was an experienced member of the paparazzi. I placed Frank in front of the window displaying the cakes and took several shots. Who'd dare to challenge me taking pictures of my husband, right?

Another interesting thing in Japan is that non-Japanese-looking people (who are very scarce) greet each other in the streets. We were looking at the map in the metro in Tokyo and a Frenchman came up to us, offered help, and even took a detour to take us to where we wanted to be. Everyone understands that it's extremely difficult to make your way around when even the slightest possibility to communicate is not there.

It took me a couple of encounters to realize that I was visually surprised when I saw familiar-looking dogs in the streets accompanying Japanese people, rather than some Japanese breeds I'd never seen before. I never thought what to expect in terms of house pets, but it did surprise me to meet a cheerful golden retriever in Tsukuba. He was better behaved than those I knew before though. I tend to talk to big dogs in the streets. In the West, dogs either respond happily or stay away. Those few dogs that I tried to befriend in Japan seemed as surprised with my familiar Russian chat as much as I was surprised to see them.

For the first time in my life, I had a total failure in obtaining something that seemed to make sense and could be paid for. We wanted to rent bicycles for everybody (five of us) for the two weeks we stayed there. The rental point was near the bus terminal (10 minutes walk from the hotel), and work was 20 minutes walk from the hotel but in the opposite direction. My idea was to rent five bicycles, pay in advance and leave a deposit (a fair business deal you would have thought), since it only made sense to have them if they were available first thing in the morning in the hotel. The rental leaflet was offering the bicycles from 9:30 a.m. until 6:00 p.m. on a daily basis. But there was no way I could negotiate this myself. I had to insist that the concierge in the hotel call the rental place. First he looked rather scared and surprised that I was trying to ask for something that was clearly not on offer on the leaflet. I explained my point to him. He pretended to understand, called the rental place and talked for quite a long time, and at the end of the conversation, turned and said, "Sorry, no." I was stunned. I was offering cash for something they were selling. But there was no overnight rate in their offer, so it was clearly not possible. Who knows what the concierge was asking them. Most likely something along the lines of 'Hey, there's this crazy woman here in front of me who thinks it's possible to do things that are not written down."

The pinnacle of orderliness came in the breakfast room of the same hotel. There were only set meals, you couldn't get buffet or à la carte (and it was claiming to be an American diner). In the menu I was ordering, I tried to ask for a piece of toasted bread instead of the pan-cake (you would have thought that the toast is less expensive). But no, it was not on the menu. Eat your pan-cake. Or if you want a piece of bread, order the second menu, eat the bread, and leave the rest. I went to a supermarket and bought a lot of sushi instead.

Thanks to Japanese astronaut Koichi Wakata, we were able to visit a specially organized tea ceremony and a Buddhist temple in Asakusa. I'll spare a description of the tea ceremony as there is plenty written about it by experts. What was special about it in our case was that we were extremely lucky with the timing. It was on the eve of the 3 February, which is a special traditional day called 'setsubun' that brings happiness to homes. The Head Priest of the Asakusa Temple, who was kindly receiving us, let us into a special area for a prayer and then offered to us all some small bags of beans. That evening each of us had to eat the number of beans equal to our age and to throw the rest away. This brought goodness in and let evil out of each of us. The joint bean-throwing event later that evening was a marvelous closure of what already felt like joy, fun and happiness.

Training at the Japan Aerospace Exploration Agency (JAXA) was as culturally enriching as visiting Japan in general. Everything was extremely well thought through and efficiently located. The test lab is next to the class-room and it's possible to have hands-on practice as soon as you learn something new.

The Japanese word for 'yes' (sounding like 'hi' in English) actually means "I acknowledge that you are talking and I'm trying to hear you." It in no way means that the listener agrees. On a couple of occasions after class, Frank would tell us with great laughter how the tutorial went. The instructors let the crew take their own course of action, saying 'hi' to every inquiry along the way but at the end saying 'No-no-no' as a conclusion to this train of thought. "What do you mean 'no'? You agreed to everything we told you along the way!"

"Hi".

Some weeks later I bumped into my friend Arie Bossche at work.

"Arie, I've been to Japan, what an amazing experience!" I said as we passed each other on the staircase rushing to meetings.

"I suppose you speak Japanese now. Hi!" he immediately nodded back, with a grotesquely humble bow and a big smile.

## Cologne, Europe

Cologne is a lovely old town with a major European cathedral. A magnificent eclectic mix of ancient and modern. As on the other occasions where I mention history and culture of the places we happened to visit during these past four years, I would wholeheartedly encourage you to read specialist books about it and visit if you get the chance. I'm not an expert in any of this, but I can tell you it's sure worth it.

The European Space Agency has its headquarters in Paris and several sites in other European countries. The European Astronaut Centre is situated near Cologne, Germany. A three-storey glass building hosts about a hundred people and a wealth of knowledge about human spaceflight. The training hall has all the facilities to train for in-flight operations on all the European elements and scientific facilities on The Station. Once a year, around the middle of September, the European Astronaut Centre takes part in the German Space Day and opens its doors to tens of thousands of visitors. If you're in the area, go and have a look. Check the website for the date and details, and if you can invest your Sunday into this special opportunity, there'll be hardly another chance to see so many astronauts together in one place at the same time.

The European Space Research and Technology Centre (ESTEC) in Noordwijk, the Netherlands, is the largest site of ESA. Space Expo, the visitor centre next to it, gives a well displayed overview of the European space vision and exploration, starting with the dreams of Jules Verne and reaching out into deep space. If you're on holiday at the Dutch seaside, there's a good chance that the local weather will give you an opportunity to visit this museum. Have a look at the Space Expo before you go for Rembrandt or Van Gogh.

# 5 Baikonur

*"If you bring forth that is within you,*
*that which is within you will save you.*
*If you don't bring forth that which is within you,*
*that which is within you will destroy you."*
*(The Gospel of St. Thomas*
*Gnostic Gospels, No. 70)*

*"The healing comes in the writing, not the reviewing of it at a later date."*
*(Dr. Joseph Mercola)*

Baikonur... Blooping hey...
Where do I start...

Noam Chomsky, who is considered to be the father of modern linguistics, designed and presented in the late 1950s a sentence with correct grammar but nonsensical content. He used it half a century ago to illustrate the need for more structured language models. The sentence reads 'Colourless green ideas sleep furiously'. A nice innocent little sentence. Unless you start thinking about a meaning, it almost kind of makes sense, doesn't it?

After a number of years of living in the West, I grew to believe that there is room for discussion when you honestly and openly address people with your problems and feelings. At the same time I derived a lot of amusement from the inability of the Westerners to see through the farce and ideologically driven role-playing. So many entities are still run in the best traditions of the blooming years of what is now called the 'stagnation' period in Soviet history. Every time I saw something that looked like a big window-dressing exercise, I could not deny myself the pleasure of sharing my observations with my husband and occasionally with a few close friends. Eventually it always turned out that I was right.

"How did you know that?" the question kept popping up.

"There are 150 million of us who know," I kept repeating, to date the only honest explanation I have.

The same as it's obvious (to about 150 million of us) what kind of power games were to start as soon as the former Eastern European countries joined the EU, it's also visible to an attentive but naked Russian eye, when people really work on content as opposed to only creating the impression of being deeply involved. There is an expression for this in Russian which can be literally translated as 'to display boiling activity'. In every conversation of this nature I've had, I use an old Soviet joke as the explanatory argument.

There is a train running on a railtrack. The railtrack ends. What would each Soviet leader do to continue the movement of the train? Lenin starts by answering that he will call 'subbotnik' (extra free work on Saturday) and build extra track in a free day. Going one by one through the solutions from the other leaders, we get to Brezhnev. "Let's close the curtains, say 'tu-tu', shake the train and we all assume that we are still driving."

I was growing up when this was an important component of the national philosophy. Millions of us were. This is how I know farce and puppet play from the truth. About 150 million of us still do.

There is another good Russian expression: 'door-guard syndrome'. I say 'door-guard', but the actual Russian word I mean does not translate very well. The meaning of it is someone who finds themselves in the position of having a choice to say 'yes' or 'no' (to let you go through the door or not, hence the expression) and they always choose to say 'no'. This gives them an immense power over the one who is trying to pass. The hapless seeker might start negotiating, and eventually begging, if they need to get to the other side. What a pleasure for the one in charge to maintain the play of not letting them go through!

Some months before the launch, it became clear that the crew had to go to Baikonur two weeks in advance. Even though I have several mutually complementary theories, I'm not going to offer any personal speculations on what possible motivations the system used to decide officially to implement this approach. The only thing I know for sure is that, for his previous flight, Frank had to make an up-and-down trip to Baikonur about two weeks before the launch in order to do some preparatory work, and then the crew was taken to Baikonur five days before the launch. According to those several cosmonauts and astronauts with whom I've had a chance to discuss this, and who flew on such a schedule, this was a very nice supportive approach that allowed all the work needed to be done ready for the launch at a comfortable pace. It took into account that extra time far from home without any real work is anything but contributing to psychological support and emotional well-being.

Frank and I have a few habits that might not be typical for a married couple. One of them is that we love spending our free time together. You might laugh reading this, but I can't tell you how many times Frank got strange looks in response to his simple and honest statements that the best psychological support he could have is to have me around! It took me a while to figure out what it is in the Russian mentality that is so desperately counter-comfort, counter-support and counter-family encouraging. I really wondered why this was happening to us. After all, it should be somewhere deep in the Russian soul that good wives follow their husbands into Siberian exile.

But that was December 1825. What happened in Russian history since then to bring forward the regulations that force people who are just about to fly in space for half a year to have to 'enjoy' another half a month of deprivation of human contact and touch? The most interesting thing is that, while on the 'working level', many people admitted in secret that they shared our views and feelings, the proponents and guardians of these rules seemed to be sincerely convinced that it was for the good of the world in general and human spaceflight in particular. Or at least they consistently and aggressively acted this way.

After a while it was clear that, despite all our requests, the only response we would get was an extended lecture about Soviet dogmas. Frank was getting more and more astonished as he heard stories about the launch and landing timelines of thirty years ago, that Korolev said that women shouldn't be allowed near a rocket, and that simply 'this wasn't the done thing'.

Basically, he hasn't once heard anything based on objective reality. Any counter-arguments he gave were met with explosive resentment, as so frequently happens in life when an opponent knows that they're wrong but can't afford to admit it. Frank couldn't quite believe how the people charged with the physical and psychological well-being of the crew plainly disregarded inputs of a crew member on what would constitute psychological support in their particular case. The truth seems to be that the system was designed to optimize operational goals and to legitimize only a small range of emotions. There's still a culture that promotes heroism in an emotional vacuum.

Frank had to leave for Baikonur. An extensive programme, which to a large extent consisted of signing envelopes and endless crew photos, was devised to keep them occupied. One of the arguments that kept being repeated to us as to why I couldn't join him was the shortage of facilities to accommodate spouses. The moment the crew arrived in their quarters and went online on the internet, I got a skype call with video from Frank. He pointed to the large bed in his room. The first thing the other guys of the crew asked him when they saw his room was: "And what was the point of not letting Lena in with you? We're used to her being around, and you for sure have a big enough bed."

No, it wasn't our intention to have me in the crew quarters. We were just asking to have a room in the building. There were clearly enough of those as well. Alas, hopefully some people will wake up and be kinder to the generations to come.

When you fly for a short mission, an extra week or two don't make much of a difference. Long business trips are the norm in the space business. A half-year trip is another matter. Extending it by half a month for no real purpose seems to go as a given with the Russians who are used to suffering in order to achieve.

As naïve as it sounds, until the night before the departure to Baikonur, I was clinging on to the hope that this universal injustice would be resolved. When people see how much we want to be together and how much additional pain and anguish this undeserved punishment brings, someone will find a solution. If you have a hope it doesn't mean that you have a chance. Come on. You're forty years old. You should've known better.

I had about ten days to spend in Moscow. Fortunately, since I was born and lived there until I was 22, I knew enough people to keep me company. It was precious, priceless and invaluable how close I became to Julia. With Brenda across the globe, it was only the two of us who knew what we were going through. We didn't need to talk about it. We just enjoyed our occasional time together.

At three days before the launch, we finally arrived in Baikonur. The infrastructure for accommodation was good. Much better than what I would've expected. Our hotel was within walking distance from the guarded quarantine area where our husbands were presumably kept well. We finally got to go to see them.

According to Wikipedia, Nim Chimpsky was a chimpanzee who was the subject of an extended study of animal language acquisition at Columbia University. He got his name as a pun on Noam Chomsky, the foremost theorist of human language structure and generative grammar at the time, who held that humans alone were 'hardwired' to develop language. What can I say? I don't have a better image to describe my experience there and remain polite. The self-rescue coming in the form of your ability to indulge in self-irony can't be overestimated. Especially when you've no one to turn to for sharing your progressive feeling of the Earth breaking open under your feet (with the digging done by those who claim to be supporting you). The nonsense we had to face was nowhere near poetic. The recognised Russian equivalent of Chompsky's nonsense sentence I used at the beginning of the chapter sounds even more nonsensical because it's made of nonexistent words (for those interested: гло́кая ку́здра ште́ко будлану́ла бо́кра и курдя́чит бокрёнка).

Creatures named by non-existing nouns perform actions described by non-existing verbs. Why all of a sudden does this sound all too close and familiar? I felt as if I'd entered a militarized zone populated with the characters from questionably successful linguistic experiments where by some odd misfortune some of my family and friends were stuck with me.

Unlike me, the others in the visiting groups had not seen 'their' crew member for much longer. Brenda had just flown over with the whole family from Houston where Bob had left about two months before. Frank's kids, Nele and Koen, had travelled into Russia to join me on the Baikonur trip also some weeks after we saw them last.

We were brought into the room for press conferences. It had a glass 'aquarium' for the crew. We were literally separated by the glass and the audience had to sit in an otherwise big normal meeting room. We were sat in front of this stupid window, given microphones and were instructed to wait. The guys entered to sit on the other side of the glass.

"You can now talk," came the instruction from somewhere.

The nice kids, with visible effort, started asking some random questions along the lines of how was the weather in the last two weeks. I had to keep my glasses on. It's good that big glasses are in fashion. They allow the covering of not only very sad eyes, but also half of the face that comes with them.

I couldn't help the thought that this meeting was even worse than in prison. There, if one believes American movies, the visitor gets a separate section of the glass wall with a personal phone-like receiver on either side, so at least you have a chance to have some kind of private conversation. But this wasn't prison. Privacy was not assumed. The conversation was artificially maintained for a while and really ran out. All those things the families wanted to share with their loved ones were not to be discussed in front of each other and the people who were there to accompany us. No matter how well they mean, they are strangers, colleagues tasked with escorting us as they would be escorting the others six months later.

"Any more questions?" came a question from whoever was organizing this treat.

"I have two questions," I said as I raised a hand in order to acquire a microphone.

"Where is the toilet? And when will this circus be over?"

"The toilet is across the hall and please shut up," whispered Frank into his microphone, as he jumped back into the real world out of his forced happy performance. "Otherwise it will never be over."

We had discussed it often enough and Frank knew only too well that I am not good at participating in farces. I wouldn't have had a problem if I was told that I wasn't going to get any help and I could forget it. Then it would've been up to me to decide what to do on the basis of the honest information I was given. What really gets me every time is when people write their own script, allocate me a role, force me into playing it and expect me to pretend that I'm enjoying it. I grew up with this. I thought the 'Cold War' and its propaganda were finished over twenty years ago. It didn't help that I understood every word said around me. People in that outfit were more or less from my generation. I knew where they were coming from in the broadest meaning of the word. I couldn't believe that I had to relive my communist past from the late 1970s and early 1980s on the eve of my husband's departure for six months in May 2009. Frank encouraged the crew to proceed to the exit.

Frank spent the ten days leading up to this final chance of togetherness trying to be totally cooperative. In his Belgian head he expected that when it was clearly seen that he was friendly and supportive, it would also be clear that he was not looking to revolutionize the system, but simply would like to be left in peace to spend some time with his wife. I was telling him over and over again, that while this might be a winning strategy in collaboration, for example, with other Belgians, or people who share these values and mentality, you don't get far doing this with the Russians. Unfortunately I was right again.

In the Russian mentality, if you show a sign of what in the West would be considered as goodwill, it will be taken as a sign of weakness. To clarify: if you begin bargaining with a Russian, you say 10, he says 5, you say 7.5, he assumes 7.5 is your next starting point and expects to settle around 6. Frank's peaceful collaboration seems to have been taken as a sign that he acknowledged that his request for having me with him was wrong, and he was happy to surrender to this wonderful and clearly the best system in the world for processing humans for flying into space.

After showing our tongues and having our temperatures taken by the quarantine doctor, we were finally allowed to face our husbands without a glass wall between us. "You are not allowed to hold hands and you are not allowed to kiss as you are walking outside," came the instruction that made my nails grow. What was the purpose of this glass performance if we could meet afterwards?

It's interesting enough to walk once on the premises of this quarantined area. According to the Russian tradition each cosmonaut, starting with Yuri Gagarin and including everyone who has ever launched with the Russians plants a tree there. These are two roads shaped as a "T-junction" alley with very tall powerful trees from the previous generation and brand new just planted trees of the recent cosmonauts.

A swimming pool drained of water in this wild garden reminded me of another classical Russian joke which felt ominously appropriate for the circumstances. In a psychiatric hospital a group of visiting professors is observing a group of patients who keep jumping into an empty pool. "Why are you doing this," they ask one of the patients. "We were promised that if we show good results they will pour water into the pool for us."

After two hours of pointless hanging around, in luckily nice weather, everyone was getting tired. We all wanted to sit down and have a cup of tea, since it was starting to get cold outside. Frank went to try to arrange tea for everyone in the conference room where we had met earlier. There was a big oval table in that room, as it happens in good conference rooms. Frank succeeded, after some struggle, to get tea for our group. "But then the crew has to sit on one side of the conference table and everybody else has to sit on the opposite side. Otherwise you might share bacteria while drinking tea."

After a day on the road, a hypocritical conference through the glass and over two hours of staring at historic trees, this was getting all too much for me. My diva sunglasses came in really handy when I was struggling to keep back the tears. Those were tears of incredible tiredness and helpless humiliation, having to suck up to some faceless person who was enjoying his power to wipe his feet on your already overstretched soul. Frank was still under the illusion that trying to cooperate might help. I put myself in a far corner of the conference room and pretended to be busy with my iPod in order to avoid having to talk to anyone. I knew that if I had to say anything about the situation, I wouldn't be able to stop for a very long time. Chris Hadfield noticed that I was on the edge of collapsing. Chris is a Canadian astronaut training as a part of the back-up crew for this mission. He came up to me.

"Hey, Lena, it's Dmitri's birthday tomorrow. What do you think, shall I give him my birthday gift first thing in the morning or shall I wait till the evening, when we might have a little party?"

Sweet dear Chris. As any intelligent person who has raised kids, he knew that the best way to calm another person down is to give them a completely different and neutrally positive subject to concentrate on.

"First thing in the morning will be cool. I'm sure he will appreciate it. Make sure to say that you want to be the first to wish him happy birthday this year, this might be a nice personal touch," I replied, confident that my voice was not trembling, but taking the sunglasses off was out of the question.

Helpless colourless green sleepless furiousness. No nonsense. Plain torture. A good plan to keep us all in shape before the launch.

Eventually we were told to leave for the hotel. I still spent some time (without sunglasses was a major achievement) talking to a few people. I had to keep a straight face. Nele and Koen are old enough to make their own choices and emotional experiences and yet I felt that it was my duty to avoid encouraging any upset feelings in them.

I was about 11 or 12 years old when, in front of the whole class, instead of just giving me a bad mark for an obviously wrong answer, my maths teacher started a sarcastic conversation about my body shape. Yes, I grew up into my adult body form maybe a year earlier than most of my friends. I can talk about it openly and easily now, almost three decades later. But can you imagine what a teenage girl feels when a dominant authority figure says with a squeamish expression in front of all her classmates, "And how many layers are you wearing under your trousers?" Until my Baikonur trip, this ancient incident was at the top of my personal 'ultimate humiliations of my life' list.

Finally, I was alone in my room. It was the right time as well. I stepped under the full blast of the shower. I would have never believed before this trip that I would find myself doing this. For me stepping into the bathtub in the evening is a pacifying, unwinding evening ritual that brings on positive sensations no matter what has happened during the day. Here I was, in a fancy hotel in Baikonur, not knowing what to do, but I had to do something to stop myself from exploding. I thought they only did this in American movies under really horrible circumstances. But there I was, standing under the shower hitting the tiled walls with my fists as strongly as I could, with liquid running out of my eyes and nose, screaming and shouting in anger and humiliation. I don't know how much time I spent like that. It felt like forever because it took forever to get out of my system what felt like emotional rape.

In the ten days I spent alone before this trip, I had a lot of unplanned time for unexpected contemplations. It became clearer than ever to me that in the Russian (inherited from Soviet) mentality, individual human life has no value, and the real value is only given to a collective achievement (with the exception of the sportspeople who win medals in world championships in individual sports). While this allowed the former USSR to build the biggest powerplants on the banks of Siberian rivers, this super-rich country still has pensioners dying of starvation and orphans living in poor and malnourished conditions. But they are superior in human spaceflight. As a group. Who cares about the feelings of individuals who work in this system?

What is important is to proclaim that they are cared for and their wishes are taken into account. Until this trip, Russian spouses were never even officially allowed to visit the launch! Some of them made it in their professional capacity, because a few of them worked one way or another in the space industry. As the story goes, a deadly accident occurred once in the distant past when some uninvited unlisted woman was taken onto some bus related to the launch. And since Korolev said that there should be no women near the rocket, clearly it was that poor woman's fault that the rocket exploded and people died. No, I'm not joking. I've heard this story more than once as a valid argument accompanied by the apologetic but firm refusal to support my obviously unforgivable criminal desire to keep my husband company during his free time.

It goes without saying that each couple should have their preferences respected. I fully appreciate if some people say that they prefer to say a quiet goodbye in Moscow and consider that the mission has started with the departure to Baikonur. If this is how they feel about it, this is how they should be supported in doing it. But if others feel differently, instead of hearing out what makes the difference for each human in this human spaceflight programme, why does the system declare them all 'supported' and push them to fit into the rules devised during the years of the Cold War in the last century!?

Russian cosmonauts get a title of the Hero of Russian Federation (in the past it was Hero of the Soviet Union) after their flights. This is the highest government reward. It was established in the mid 1930s specially to honour those who displayed acts of heroism in military action and to those exceptional civilians who contributed in unique ways to the field of aviation. In some moment of clarity, this whole puzzle of cruelty started making sense to me. You can't be a hero just because you've done something very well. In the Russian mentality, heroism goes with pain and sacrifice. Thank God there's no war in our part of the world (too bad there are one or two in the other parts). The whole environment surrounding their attitude to the family in Russian spaceflight seems to increase the possibility to sacrifice and thus to become a real hero. The system is built in such a way that it reinforces this opportunity to sacrifice, and leave behind the dearest things you have – your family. Bingo! A Procrustean bed of personal, family and psychological support. When would they wake up in the 21st century?!

Two days before the launch, the traditional ceremony of tree-planting took place. It was jolly nice and well organized. The upper part of the T-junction valley has acquired three new trees.

The crew was escorted to and from the ceremony. The set-up was well done. Three shovels and three buckets of water were waiting for them, lined up behind the rope that was meant to stop the press from crossing the visible divide. The evening before, one of the camera crew got sick. He went to a hospital (just in case, as far as I understood) but because of that one of the Belgian TV stations didn't get video coverage. This made it into the major evening news. After 7:00 p.m. Belgian time, I received a few text messages with desperate questions about whether Frank's swine flu was now confirmed and the mission was cancelled. Gosh, our human nature and this insatiable craving for a major drama!

When the crew left the planting scene, I stayed there with Frank's friend Jack, Koen and Nele to take more pictures of our own. The cotton rope divide didn't play a role anymore after the crew was gone. The kids went on both sides of the tree and leaned on the sign with their father's name on it. Click-click-click. History recorded. "Lena, come, let's take some different pictures!" I happily stood between them and behind the name sign. As I put my arms around their shoulders, I felt that something horrific was happening.

I have to tell you here that the planting itself was carried out in good conventional gardening terms. They put the short tree stubs into the holes that were already dug, filled them up with the soil that was already lying next to the holes, poured each bucket of water into their respective holes and stepped onto the filled area with their heavy boots to help the wet soil settle.

As I was standing with my arms around Koen and Nele, I felt that my lacquered boots with pointy toes and with flat but rather high heels were slowly sinking into the swamp-like soil that only minutes ago was set in by the heavy flat boots of my husband. In less than a second I knew that I was committing an unintentional but terrible blasphemy. On one hand, it actually was hilarious, but on the other I was completely horrified by what was happening. Some camera crews were still there and eagerly pointed their cameras at me. I think those were only Russian camera crews. They did not find my accidental drowning in my husband's tree hole interesting. Or maybe they heard my plea for help. Either way, thankfully, I didn't make it into the news. "Space Mission in Jeopardy: Wife Destroys De Winne's Tree Day Before Launch" or anything equally exciting.

I had to straighten the tree and to readjust the soil to make sure it remained upright and grew as intended. All the tools had been taken away by then. None of us had big shoes, and I was the only one covered in mud. Unlike the crew who operated with shovels and heavy boots, I fixed this breathtaking mistake with my bare freshly manicured hands. I wiped off what dirt I could on the grass. By the time we made it back to their building, most of the mud had dried off but I still desperately needed to wash my hands and shoes.

One of the guys in charge of our humiliating tea ceremony the night before was standing near the building where we were meeting our husbands. He offered me use of the sink in his office and offered some cloths to wash the shoes properly. But then he said, "Lena, while you are here, I want to apologise for yesterday. I had to do it because of how things are here."

It wasn't in my plans to discuss my experience of being there or to talk to anyone apart from my family or the people who were there from our side supporting us. But I couldn't hold myself back when I heard this address. I don't remember the details of what I was

FRANK, BOB AND ROMAN IN BAIKONUR

saying. Probably it was something along the lines of what you have read above. He found a pause when I was getting more air.

"Lena, please take as much time as you need, feel free to use my office if you like. I'm sorry." And he stepped out of his office leaving me a clean white crispy towel for wiping my black muddy shoes.

In the afternoon, we were forced again to walk outside in case we wanted to spend any time together. By the end of this day, Frank started to realize that no matter how friendly his cooperation would be, no one on the Russian side would make any move to let go of this hypocritical scout camp where married middle-aged couples and their immediate families can't have comfortable time together. The best we were allowed, if not to walk outside, was to sit in the cafeteria on wooden stools.

After a sleepless night, my body was becoming oversaturated by drowning in mud, endless walking up and down the tree alley and sitting on wooden stools. My body was begging for rest. The wooden stool didn't do it for me anymore. I wasn't able to sit straight, let alone walk. There were couches in the halls of the quarantine hotel. Frank had to do something for about an hour. I did the only thing I could think of. I laid down, closed my eyes and decided that in the unlikely case anyone dares to talk to me, I would look for a solution only then. For the rest, it was their problem. So what if they weren't used to strange women sleeping on their couches in the middle of the day. As is often in life, non-invasive but daring approaches work. It was funny to half-hear Russian conversation of the people who were discussing who I was and what I was doing there. They didn't know that I was Frank's wife. They didn't realize that I was a Russian speaker. My mind was drifting between sleep and reality, but at least I derived some amusement from observing how the external world reacted to the static disturbance I was presenting.

On the day before the launch, there were some interesting ceremonies to follow. The Rocket roll-out is impressive in that you would have never expected that you can stand next to it as it rides by. It's a tradition to put coins on the railtrack. The coins get flattened by the train and then you keep them forever as a souvenir. I'm carrying one now permanently in my purse. It's also a tradition to collect rusty railtrack nails. They are about as long as the palm of your hand and have at least a one centimetre cross-section. There are plenty of them lying around there. One of them is now lying in my private space debris collection.

It was extremely windy. It was almost impossible to believe that the early morning temperatures and wind, in which you would want to wear a good jacket, will turn only a couple of hours later into the heat we will all look to escape from. We were also taken to watch the raising of The Rocket. Unlike the American Shuttle and the European Ariane rocket, which are rolled out in their vertical launch positions, the Russian Soyuz Rocket is rolled out flat and raised into its launch position at the launch pad. It was still terribly windy. The wind caused a delay in this procedure. I didn't get to see it because this would have meant staying longer in the area and I wanted to get back to the quarantine quarters as soon as Frank was given free time.

The most ironic thing was that the back-up crew, who were undergoing the complete pre-launch sequence and lived in quarantine quarters together with the main crew, were out there with us to watch the roll-out. After this tour, they went back into the quarantine.

The theory was that the quarantine was there to isolate the crew who were just about to fly from any possible contact with external germs. Once again, clearer than ever, the fake and hypocritical set-up was grinning straight into my face. I had to be told when I could or could not as much as hold hands with my husband, while the people who shared their isolation were allowed in and out of it the day before the launch.

Back to our personal torments. By the evening of the second day, it was clear that the farce would never end. We resorted to something that I can laugh about now, but I was swallowing tears of helpless humiliation again when I was going through this. With care and secrecy, checking the routes and the staircases in order to avoid those people whose job presumably was to take care of the crew's physical and emotional well-being, we made our way to the room of the person who was there to support Frank (just to be clear, no one in the whole training process had caused us so much grief as those glorified caretakers on the Russian side). We managed to find some delight in the fact that we were forced to behave like teenagers trying to keep their love affair secret. It was amusing to be helped in doing so by well-paid and highly trained professionals.

For a Shuttle launch, there is a clear procedure that includes various personal activities and gives a dedicated location for the families of the crews to spend time in private. All the families are treated equally, and there is a complete transparency on what and how this is done. But why should anyone on the Russian side care that there is a valid approach that works and learn from the experience of their partners, right?

Yes, a couple of private hours in the two days before the launch were smuggled into our life, thanks to European creativity.

The day before the launch I bumped into a big Russian boss, who I happened to know from before. "Lena, how are you doing? I hope you are well," he enquired.

"Well, now that you're asking, may I tell you how I'm really doing?" I asked in return.

"Hmm I've heard that there were some problems," he said.

"Let me put it this way. My husband and I had to beg for a permission to be given some time alone by such and such. The last time I was so humiliated I was 12."

We were speaking Russian. I didn't say 'time alone', I said exactly what I meant. He was taken aback by the firm clarity of my expression. I hope I managed to express politely but convincingly how it was exactly for us.

"Come on, it's not that bad, is it?" he offered and it was clear that he wasn't sure what to say. It was that bad. It was exactly as bad as I'm describing it.

Baikonur. Blooping hey. Russian ways, my historic motherland. I thought Russian parents had long grown away from the belief that punishment brings out the best in people. What did I know.

# 4 The launch according to…

## …Roman

As the story goes, no matter how much you're looking forward to your first flight, there is always a chance your legs will be shaking really badly when you have to walk up the stairs into Your Rocket. This is probably why every crew member has two people walking with him to The Rocket, holding him under the arms. I wonder if it has ever really happened that somebody's legs gave in. From the footage I have seen some walk at the same pace, for the others it looks as if they are almost dragged by the accompanying people, and yet for some others they can barely catch up with the almost-running cosmonaut who is so eager to get into space. I will have to watch the footage of my own walk later. I think I just walked. I wonder how it looked to everybody else.

The Rocket – My Rocket – Our Rocket is there. Steaming. Like a racehorse waiting to run. Shaking in excited anticipation. Steam coming out of its nostrils. Or its nozzles. Almost the same.

The ladder up to The Rocket is said to be steep. Hence the tradition – you get a kick or a shove (well, a slap these days) on your bum. It was supposed to help you climb up the first step in the suit that makes you rather immobile. The first really helpful 'kick' to have been made must have been a knee held up for a few seconds so that you could lean on it. Nowadays it's just really a slap. I think I was slapped with a hand, and not shoved with a knee of one of the big bosses present. I have no idea who it was. I will have to see that on the flight footage after my return too.

In any case, this was a good point to break off and start moving up and saying a proper goodbye. Before that it was a bit cloudy for pictures and hand-waving (no handshakes, we are supposed to be bacteria-free). Even though the site had very limited access, there were still a lot of people there.

This slap-dash-kick was indeed a good point for another kind of break. I got a bit carried away, further than I think necessary, before the launch into sentimental realms. There is another tradition of making a movie for the bus ride to The Rocket for the crew members to watch. This movie is from friends and family and is meant to give you support and make you feel better. I think we have something to learn from the West here – all the clips dedicated to Frank and Bob were cheerful and playful. They were personal, but amusing. Russian people have this extreme heaviness in them for saying goodbye. No matter where you go, there is a drama of upcoming loss stemming from any separation. It makes any departure feel as if it is for a seriously bad reason. I know it's traditional and everything. I'm all for traditions. I just wonder would there ever be any room in the Russian soul to celebrate more than to commiserate and dramatize on such occasions? Don't get me wrong, it was a great movie, I'm grateful and impressed with all the work. But it might have been easier to watch something like that a few weeks later. You see, I'm also Russian, so I resonate strongly with separations and goodbyes; especially with those that are intentionally strongly emphasized.

It takes at least 30 minutes between the most important slap on the bum of my life and being finally strapped into my seat. Waving from the ladder, pictures and smiles for the cameras. Only very top management and press are around us. Too bad our wives are not allowed to be there. Julia means everything to me. She is the person to whom I would have loved to make the final wave before my trip.

The real sense of weightlessness came after one and a half hours when I could release the straps. I just floated up in the position in which I was sitting. Mind you, I'm not sure if it counts as 'up' anymore. In relation to where I was sitting down, it still does. It's the same as in parabolic flights when you wait for the 'drop' part of the parabola to come. First you hold onto the rail on the wall and next thing you know everything lifts into the air. This time we all released our belts at different times so I floated up alone. I had waited for this for 11 years.

I took a little toy from my daughter Nastya as a 'weightlessness gauge'. It was a character from a popular Russian cartoon, Smeshariki (literally 'little laughters'). A few minutes into the flight, this 'little laughter' had loosened up and started freely

floating in the cabin in all directions restricted only by the string by which it was attached. And then, like everyone promises, you literally see your dashboard lifting up. I was expecting this, but still it looked unreal and a bit scary – is everything all right with it? And a few minutes after, weightlessness caused a very strange sensation in my head that soon became a trivial headache. We were warned that this might happen. Forewarned means forearmed, I think this is how the saying goes.

To be entirely honest with you, the launch and the first day of the flight didn't feel like a big deal. Compliments to the trainers – everything is exactly as in the Soyuz training mock-up. I finally plugged into what was happening to me when we had to go to sleep, and I saw Bob and Frank hanging upside down in some random positions, impossible to imagine on the ground. I tried to do the same but it didn't quite work. Finally I found a way to fix myself without disturbing the others: I put my head against one window and my legs against the opposite window; and this way for the first time I fell asleep in space. I woke up every hour. Maybe the new and unfamiliar bed was too hard?

Our communication to the ground during the flight was not so good. Occasionally we could hear them, but they weren't hearing us very well and vice versa. Frank was shouting most of the time when he had to talk. That is when the ground was not hearing him. It didn't help: we were a bit too far from the Mission Control. If you think about it, it's clear that shouting won't help but this is a natural human reaction to raise your voice when the listening party keeps saying, "I can't hear you." In the last part of the approach The Station crew was providing communication coverage for us. We could hear Frank's voice via The Station, when he was shouting to the ground!

# ...Julia

The Rocket was extremely quiet for a long while. I knew that it was supposed to launch at 4:36 p.m. local time and yet it caught me off guard when everything around us started shaking. I didn't know what hit my senses first: whether I saw the flames or if I heard the noise or I felt the vibration through the floor. This vibration was coming up my spine and made each hair move individually. My face probably reflected an innate horror. A few people, at the same time interrupting each other, started telling me that this was all fine and this is how it was supposed to be.

Before this moment I could watch our husbands' faces on the monitors. I was invited to join the group at TSINKI – the location where local launch controls are hosted and where the top management spends some of their pre-launch time. I arrived about 20 minutes before the launch and couldn't take my eyes off the monitor. I found it astounding how calm the guys were. So many people on Earth are going crazy and losing sleep over this, and they are just chit-chatting, lying comfortably strapped in their individually made seats.

The monitor view lasted for about nine minutes after the liftoff. Afterwards the line breaks off. It crossed my mind if it was broken off on purpose, maybe some secret operations start afterwards? I will never know. Their immense calmness and concentration was setting the pace to all the interactions. The Head of Roskosmos asked them how they were doing with the g-force; if it was like being in a fighter jet. Roman answered that he felt it as one and a half times heavier, and then referred the conversation over to Frank. Frank thought it was twice as heavy as in his fighter jet. Unlike me, they were totally in harmony with what was happening, just chatted about the usual boys' things, probably enjoying the fact that they were finally there.

Only when I watched the recordings of the launch later did I notice how the 'little laughter' of Nastya swung in microgravity. I'm not sure she entirely understands where her father is. Roman travelled a lot before, but in her lifetime he was never away for more than three or four weeks at a time. And of course he always came home with a lot of presents for the kids. We spoke to her a few times before the departure to Baikonur. We tried to explain that this time he will be away for a few months, but we will all stay in touch in various ways. Roman's father, cosmonaut Yuri Romanenko, helped us a lot in talking to her about space travel and this is what fathers do. She nodded every time. Until during the official pre-departure breakfast ceremony, she turned to him and asked, "Daddy, are you going to be back in two or in three weeks?" Roman is brilliant in joking his way through life. He said something which made her smile back at him. I heard it but I was not hearing the words. I'm just an ordinary mother. My heart sank when I heard her question. Luckily kids get easily distracted and forget. Too bad mothers don't.

Our son Maxim was there with me. He looked cool. He looked about and got to know Bob's and Frank's kids. And yet I knew that this was all intentional behavior, while in his own heart and soul he was hiding deep inside himself. I tried not to disturb him. I think he was doing great for a 16-year old.

I entered the empty room where Roman and Frank had lived for the last two weeks. Unfortunately Lena and Brenda were on a different programme. I had to face this emptiness alone. They would have been the only two other people in the world whose presence would have helped me to feel only one third of the heaviness of this empty space. I felt deflated. Like an air balloon. I was not a pierced air balloon, I did not feel exploded. I just felt completely and gradually, slowly but surely deflated. Things were mostly packed but not quite all. I finished packing Roman's bags and the guys took them from me for dropping into the plane.

We then went into the cafeteria where the crew ate for the last two weeks, sat at their table, had dinner. One of the doctors who worked in supporting the launches over the last 25 years could not stop repeating the story that he had seen every launch, and this one was the most beautiful ever. Perfect weather, perfect sky and perfect visibility. Perfect crew and perfect wives, I thought to myself. I share the general stance with Lena that we all have done something right in our lives to deserve each other in this most unusual and very emotionally challenging experience. So why not assume that the weather resonated with our energy as well? Lena says Frank is known for getting good weather on important days and good parking spots when he is in a hurry.

The plane was departing at 9:00 p.m. We had plenty of time to spend in this all of a sudden empty building. The others seemed pleasantly excited and spoke loudly and laughed and toasted a lot. It probably seemed desert-empty only to me. "Everything is fine," people kept telling me when they saw that I was not communicating much. I knew that. That everything was fine for now. But there are now about two days to go until the docking. I'm looking forward to see Roman after the hatch-opening. I think I will be able to breathe out then. Counter to any logic of physics, this might help me inflate again.

I was part of the group travelling there and back from Star City, which has the Chkalovsky military airport nearby to support its work. Valeri Chkalov was a legendary Soviet military test pilot who was the first to make a non-stop flight over the North Pole from Moscow to Vancouver. A few places around Star City (which is called after the first cosmonaut Yuri Gagarin) carry his name.

We landed at Chkalovsky at around one in the morning. This is a closed airport. There is a car park close to the exit, where the cars of the big bosses are standing waiting with the drivers to pick them up. The queue was long – there were a lot of foreigners on board. When I finally exited the airport, I could not believe the view. There was a small picnic laid out on pretty much every bonnet or every boot of every car! I was travelling together with my father-in-law, who of course was an honorary guest at every location. Thanking the Creator for supplying my body with a good share of adrenalin, I went on for another hour with the celebratory drinks in this middle-of-the-night picnic, where top-ranking people and flight mechanics were joined together in one continuous cheer. Our car first took us to his place. My mother-in-law was expecting us there with a laid table. I was home by four. Nastya was sleeping at my mother's place. I wonder if she understands where Roman is. That was the end of one of the longest days in my life.

# ...Bob

The last time I had so much attention focused on me was when I was getting married to Brenda. On the day of my launch I really felt like a groom on the morning of our wedding day. There were three big differences though. One was that I was doing it in the company of two good friends who were not just helping me but doing it together with me. Two, we were wiped with alcohol all over our bodies for sterilization in order not to bring any bacteria with us into the Soyuz and eventually into The Station. On our wedding day, alcohol was used differently. Three, we were dressed into some plastic-like underclothes. My wedding suit was a classic black. This preparation, more than the flight itself, gave me a feeling of being in a deep-space science fiction movie.

But this was just a start of this most surreal day. This is not a figure of speech. If I pause and quietly think about it for a moment, I would still say that this has been the most surreal day in my entire life. Not only because of what was going to happen, but because The System succeeded in making the pace of time change.

The Russian system is clockwork. Every activity of the day, including the launch itself, is scheduled down to a second. The people who run it make sure that those minutes and seconds are respected. It seems that they purposely select time like 10:37:23 for the start of the next action and make sure that they do it exactly on time.

Sit. Wait.

Pause in the schedule. It goes slow. Somebody is watching the clock. The second hand reaches the expected point.

RUSSIAN PRIEST BLESSES THE CREW ON THEIR WAY TO THE LAUNCH

Everybody up! Bob, don't fall behind your crew mates. Go quickly. Your next activity is now.

Feeling of rush and anxiety in fear of being late. Rush-rush-rush. Phew. Done on time. Stop. Stand. Pause. Anxiety-driven schedule. What an interesting concept. A surreal one.

Baikonur itself is nothing like you would ever see. It is isolated, hard to get to, impossible to access without clearances that take months to acquire. We had to stay there two weeks before the launch. In this remote regimented world we were kept busy: we raised flags, posed for the pictures, signed photographs and visited a museum. The only thing we did that brought us closer to The Launch was one fit check and visit to The Rocket. Being at the assembly factory gave me something that in psychology they call a 'closure'. After years of training in simulators, having watched numerous launches on a big screen and having supported one Baikonur launch as a back-up, I have experienced the days leading up to my own launch as something altogether different.

The day started with the State Committee where we reported that we were ready for the flight. Unfortunately Brenda and Lena were not allowed to attend. But it got much better the next day. There was a slow surreal walk back to the commander's room for a drink. All the spouses could join us for that drink, and eventually the kids and the other family members were admitted to the steps that followed. Speeches, walk out, rush-rush-rush, sign the door of the room where you stayed. My good friend Marcos Pontes, the Brazilian astronaut, stayed in the same room as me before his launch in spring 2006. I made sure to find his signature and signed next to him.

I'm a religious person. Even though I'm not a Christian Orthodox, a blessing we all got from the Russian church gave warmth to my heart and added yet another dimension to my inner belief in our success. It gave me peace and pleasure to know that religious significance is attached in Russia to space exploration. My kids found it jolly amusing though when they saw streams of wholly water running down my face. Rush-rush-rush.

The pre-launch process in Baikonur is one big complex tradition. It's a special unique experience to be part of it. Even though I have to admit, occasionally it got in the way.

We walked down stairs, along the lines formed by the people who came to watch us leave. I tried to look into the crowd, to greet and touch hands on the way with the people I knew. Not supposed to do that. I touched them in secret when I could. Rush-rush-rush.

One of the NASA people (NASA took excellent care of my family) made sure that my mother stood right in front at the bus where we were boarding to go to the launch pad. They knew how important it would be for her to hug me right before my departure. I know it broke her heart when the quarantine doctor stepped in front of me with his usual Russian 'нет' – no – and stopped her reached out arms from touching me. What could I do? I understand that he was watched by a big bunch of bosses who probably would have fired him if he didn't do it. But I really wonder if all those people can't understand that, while rules are important, human kindness exercised within the boundaries of common sense and good cooperation is even more important when you work with people?

My mother Eva watched my first launch together with my father. Those two did everything together. Now she's alone. She travelled from Calgary to Moscow and then to Baikonur for five days in total, together with my brother Rich who went really far out of his way to make this trip and my sister Bev who is Canada's number one space fan. My Mom is an upbeat person. She uses a walking stick to walk. I know she has a lot of pain and discomforts. She always smiles, she never complains. I just couldn't believe that I was made to wait until sometime next year to hug her. You might have noticed we live in different countries. Kafka-like surreal. Rush-rush-rush.

I'm really grateful they all came. The kids developed some kind of teasing way of dealing with my lengthy trips. Up to three months ago they were still discussing that it would be tiring and unnecessary for them to go to Baikonur, and that they could watch it on a big screen in Canada. I never insisted on their travel. I'm happy that they themselves changed their minds. Being with the people you know who love you is a tremendous support. It gives me so much peace and relaxation to spend time with them.

A long bus-ride with videos from our families and friends. Slow. With a traditional – unrushed – 'Gagarin stop'. That means we had to get out of the bus a few hundred metres before the pre-launch building and pee on the back right wheel of the bus.

Rush-rush-rush into the pre-launch building. Sit. Slow. Clothes change. Scrub. Rush. ECG measurement bands across the chest. Long white flight underwear. Rush. Slowly enter the 'conference room'. We are in an aquarium. The rest, in a bigger aquarium. I never understood the purpose of putting the spacesuit on in front of that gathering. This is a very refined technical operation. I would have much preferred to have it done in a quiet atmosphere together with the technician who was helping me. But OK, it provided entertainment for everybody else.

I wanted to take with me on board my crew notebook. I was hoping to put it into my pocket. My crew surgeon Doug was holding it in his hands the whole time. Formally, I wasn't supposed to do that (I never understood why because it contained all the useful notes I wanted to have for my work). I didn't succeed. We weren't alone for a single second. I didn't get a chance to take it from him discretely.

In my mind I called that aquarium event a 'greetings conference'. A lot of people came there to say hi and bye to us. Thankfully the families were invited and got good seats. I tried to find all the people I knew in the crowd, and make eye contact and wave at each and every one of them individually. This was the only way for me to say a personal thank you to many people whose work made all the difference in this experience for me and my family. I think NASA provided incredibly good support to us and our families, and that is bearing in mind that there was no NASA astronaut in this launch.

No friends or family at the launch pad. Even my crew surgeon Doug was not allowed there. I would have loved to have Brenda and Doug there with me. I was much luckier than Frank and Roman though. Steve MacLean, the President of the Canadian Space Agency, who was there in his professional capacity, also happens to be a good personal friend of mine. We were selected and started our astronaut training together long ago. We shared an office for almost ten years. It made a difference for me that he was there.

The Launch Vehicle. Finally. Stairs to The Rocket. Roman and Frank offered me to go first. We were instructed to turn around for photos. I would have loved to wave to Brenda from those stairs. Elevator, a couple of technicians to help us on the way. Finally, freedom! Me and my buddies are left alone to do the job. I realized that throughout this half fast-forwarded and half paused day, I was longing to get into The Rocket and to be left alone with my crew. The leak-check is finished. We are rid of everybody! I have full confidence in Roman and Frank to do the job and get us safely to the ISS. We didn't need all those dignitaries to keep telling us what to do.

Frank entered the capsule first. Me and Roman were casually chatting as we were waiting for him to be ready. My turn next. Slow. My own real slow. No room for errors. 'Do it exactly as in training but just in case, be slow'. There was plenty of 'pause' time allocated in this part of the time-controlled schedule: we had two and a half hours to strap in. Frank guided my legs to make it easier for me to enter. Then Roman joined.

There is an unwritten astronaut motto: 'Do not become famous.' This comes from the fact that, even though astronauts draw a lot of attention and the public tends to attribute them with some non-existent advantages and fictional superpowers, in reality we are forgotten as individuals soon after the flight is over and media attention switches to the next space traveler. The only way to 'remain famous' is to do something wrong. Hence: 'do not become famous'. I tend to translate it for myself into 'think twice before doing anything'.

Our conversation remained light-hearted but serious. We had plenty of time to prepare our documentation, turn pages together and check ourselves on each step of the way. Our instructor Kostya from Star City guided us via the comms link. I found this very nice. The same person who taught you to do the work is now supporting you in doing it. Much better than someone you have never met from Mission Control. This was somehow pacifying. The only slight disappointment I had was that we didn't get any music played to us. I gave some disks a couple of weeks ago when it was offered to provide some music that we wanted to hear while we were waiting for the launch. The quiet time went slowly and then the ultimate pre-launch 15 minutes made the final rush-rush-rush for the day. I was watching the countdown timer, but I swear the seconds were running faster than usual. Five, four, three, two, one. Our liftoff felt as if we were taken up in the hands of the angel from our Soyuz logo. Inside The Rocket, you feel the vibrations and the noise much less than when you watch a launch. And then the fairing blew off. A stream of light came my way.

I knew Brenda was somewhere still not that far away, hugging the kids, holding us all together and guarding my reality. Brenda. My oasis of sanity in this surreal world. Or was 'OasISS' the ESA name of Frank's mission?

# ...Brenda

Dali or Whistler? Johann Sebastian Bach or Dave Brubeck? Kafka or Tolstoi? La Traviata or Cats? Swan Lake or the River Dance?

Real or surreal?

In curious amusement I observed this unexpected sequence of free associations coming out of the collective consciousness straight into my oscillating reality, which kept swinging between opposite extremes since I had left our home in Houston. I knew in advance that this trip would culminate in a painful dichotomy. Sweet sorrows by Shakespeare. I'm so happy for my husband who will achieve what he had been working towards for a quarter of a century. My heart breaks when I think that he and his two friends will be sitting in a tiny shell on top of many tonnes of explosives in order to travel to a futuristic destination (I pray to God every day for its safety), which will keep him away from home for yet another half a year.

I only could travel to the launch a few days in advance. By that time, Bob and I will not have seen each other for almost two months. With all the family from Bob's side and our kids, I also had a piece of my heart with my parents who were in Montreal. They were very poor Italian immigrants who survived the Second World War and the Nazi invasion of their small town in Italy. Their formal education ended in the 5th grade. They raised two daughters in French Canada without ever speaking a word of English to them. Even though I felt totally consumed by what was happening in the here and now, my mind was also wandering off to join them as they witnessed their son-in-law's Soyuz launch on a big screen in Montreal.

The barren landscape of the Baikonur desert set the tone for what was about to happen. The sand beneath my feet, the weeds and even the bug that tried to reclaim the spot where I had chosen to stand, were all reminders that my mind was still wandering in search of harmony between respect for the natural order of life and forging ahead with innovation and exploration. Those two don't need to be mutually exclusive, but balance is critical to the preservation of both.

Earlier today, Julia, Lena and I had an opportunity to be part to the traditional drink in the cosmonauts' room before their departure to The Rocket. This felt very special and important to be made part of the traditions. The connection that developed and deepened in Russia among Bob, Frank, Roman, Julia, Lena and myself felt unique and powerful. I was seamlessly drawn into its vortex only a few days before and now, on launch day, the six of us shared far more than a mission and personal dreams. Julia and I don't even have a language in common, and yet, I feel as close to her as I do to Lena. We just can't talk directly, but somehow we all feel that we are holding hands to support each other while we are going through this. Although I spent no time in Russia before the launch trip, I sensed all along that I would have been welcome and that I was integrated and an important part of this close and delicate bond which has grown between all of us. That made all the difference for me.

I know a lot of other astronaut wives. I have heard different stories of how the training and the flight affected the relationships. This team spirit of 'our crew' no doubt began on the day that the crew was selected. Their joint desire and decision to take extra steps to include spouses as true partners in sharing their training and flight experience have shown to me how visionary, mature and committed they were. If only our micro-example could inspire broader groups of people to value human connection and relationships before political goals and monetary worth!

The Rocket seemed to understand how much we were all relying on her as she sat quietly and discretely in the desert sun. I thought of Lena and Julia and each of our families as we watched from separate locations. I reached out to them with my heart and I somehow knew that they were all going to be fine. And I knew that Bob and his mates were connected to us even though they were concentrated on doing their work in this moment. Physical separation is just an inconvenience of this otherwise shared experience for Bob and I. The desire to be near loved ones is universal. The ability to support each other is invaluable.

With ignition, there was a distant boom and flames dictated a liftoff. A massive roar spread across the steppe. The same as when I watched Bob launched for the first time on the Space Shuttle over ten years ago, I felt my core pull me into myself where I could focus and find that clarity that I needed on those few occasions in my life when something vital was going to happen. I wrapped my arms around my three children and everything but The Rocket vanished. When I turned to face my youngest, the tears on his cheeks sent a piercing reminder of the weight that would not as easily disappear. We all agreed many times before that it would be a joyous occasion, yet when I hugged him and looked into his eyes, my heart had to reassure him, once more, that his Dad would come back. I then hugged the other two and I could no longer hide my own tears of joy and of gratitude and, yes, of the bittersweet knowledge that the next six months would be the height of both achievement and longing.

I was in awe. And there was no more dichotomy left there. I was truly happy. I felt the weight of years of hoping, waiting and training lift with the thrust of The Rocket. They have launched safely. Bob was exactly where he wanted to be: in The Rocket flying into outer space with two true friends.

*But if the while I think on thee, dear friend,*
*All losses are restored and sorrows end.*
*(Shakespeare, Sonnet 30)*

# ...Frank

I have been through this insignificant but current dilemma once before. Do I, a grown man, a general in the Belgian armed forces and viscount of the kingdom, wear a diaper? I know you're laughing but there's more to it than an entertaining intro. The choice is between a seriously uncomfortable wrap around your lower body, that may or may not come handy for a physiological function in the next few hours, or being much more comfortably dressed and relying on your body to support you as it always does. Additional physical discomfort is not something to neglect when you're doing a complex job. Both options have pros and cons. When I give good thought to planning for things to work out, they mostly work out my way. There was no diaper the first time around. Let's try to use it this time.

We had the traditional drink in the room of the commander in the quarantine profilactorium, third floor, where we had been living in isolation for the last two weeks. We, the crew, were extremely happy that our wives were allowed to join this drink, which is otherwise closed from the rest of the world. Our wives go through so much for us. It feels so natural and so right that they should be welcome to share with us all the human experiences of this incredible adventure. A few short speeches. Traditional in any Star City celebration is the triple hurrah, "Ura-Ura-Ura!" after each speech. Sit down, anywhere, even on the floor, after the closing toast (another Russian tradition, all sit down together as the pre-departure ritual for any trip in order to make sure the trip goes well).

You can't say that word ('last') in the Star City Russian language. Russians are full of superstitions. It seems that saying 'last' might draw some bad luck and make it the final 'last'. Instead of 'last' ('poslednii'), the Star City word used is 'krainij', which means 'the one on the margin' or 'on the edge'. It doesn't sound right in English, so I'll use the word 'closing' in place of 'last'. This way, I share with you this small but important component of our reality and hopefully make it sound comprehensible. The inhabitants of Star City apply this colloquial rule to everything they talk about. For example, the final trip to Houston before the launch is the 'closing trip' and under no circumstances the 'last trip', lest it otherwise turn out to be the 'last ever in your life' trip to Houston. Not that Houston is a location you should keep revisiting, but there should at least be some post-flight briefings there, so you wouldn't want to call the 'closing trip' the last. That also goes for the 'closing toast' at any party, which should not be called last, in case it turns out to be your last drinking event. Even my super-Russian-philologist-linguist wife Lena gave in and started saying 'krainij' in conversations with Star City people. Mind you, every time it happened she secretly whispered into my ear a reminder that this is incorrect Russian and I shouldn't learn intentional mistakes.

Everyone out! Smash the commander's glass against the wall. Successful completion of the most important pre-launch operation is achieved!

As we walked downstairs, we received the blessing of the launching crew by a Russian priest, traditional since the fall of communism and generously accompanied by a cross-shaped splash of holy water in our faces. My crew surgeon gives me a gauze tissue. I gratefully wipe my face; give the tissue to Lena to keep. It would be dry very fast, but hopefully will retain at least some of its holiness. This cloth will find a very special place in our loft full of boxes with space memorabilia.

Flying into space is first of all a technical job for me. Of course I would be lying if I said that the attention does not flatter me. I'm a regular person. I react in a regular human way when I get positive attention. Like every human being, it taps into my ego to see all the people travel far away to see me and try to talk to me, to share my experience, to support me, to wish me a good trip. But I realize that only my family and very close friends are there for me as a person. They would also travel and see me and support me when I'm not doing anything exciting or spectacular in public terms, such as when I just sit in my backyard and listen to music or sit along the canal and fish. Or if I make mistakes or do something plain stupid. They will come and sit with me. Lena's mother likes to accompany me on my fishing trips. She seems to have the most patience of all to sit alongside me and read a book. She is tiny. My fishing bag is almost as big as she is. She's a real Russian mother – she tries to help me carry it.

It is a special honour that His Royal Highness the Duke of Brabant, Prince Philippe, came to see the launch this time. He had been to see the landing of my previous flight. He takes a special interest in science and technology. I hope this would bring more interest to youngsters to study science and engineering and more motivation in society (for example, to politicians) to invest in technological innovation.

The rest of the crowd is there for the event. Of course 'my' part of the crowd is predominantly Belgian. It would be a crowd from another country when another European astronaut is launched. This is something we should still work on – to get over the de facto existing cultural barriers in Europe. Interest in a flight of any particular European astronaut seems to come only from the country of their origin. I hope one day soon that we will be able to feel a shared pride in any achievement by any European citizen, in any area of life. I know it's difficult when the mass consciousness is alert to sports results as an identifier of national success. Shall we try to remember that sport is a game, which is why you 'play' sports, and not 'work' sports. I can't tell you enough how passionate I am about a unified Europe, its common future and shared vision. The only way anyone can win, no matter what area of life you're talking about, is if there is a joint will to look for a win-win-win solution.

I am proud that my years and years of training have earned me the right to represent the collective achievement of all the people who made it possible. It's a functional part of my Job to symbolize this collective achievement. My Job is important. But my Life is about the people who are important to me. My Life is bigger than my Job.

We take the closing tour – The Bus ride to The Rocket. I leave it to Roman to tell the story about the traditional movie that was played for us on The Bus. It featured his five-year old daughter Nastya, dressed in an old cosmonaut launch suit (I think it was his father's launch suit!), telling us a little poem that actually made me look away and pretend that I wasn't reacting emotionally to it.

The Rocket. At this point I know Lena would say that you need to develop an inner harmonious relationship with an object upon which you rely. She insisted that The Rocket, The Station, The Vehicle and so on are all written with capital initial letters in this book (or The Book I should say). In this way you show respect and appreciation to the energy and the forces you are working with. I find this sweet but I have no idea what she means. I just know that I know how The Vehicle works and therefore I know how to make it fly.

We decided as a crew not to use window shades. These shades are provided in order to avoid visibility through the side windows after the fairing is jettisoned. When you just sit in the capsule, you have no way to look outside. But a few minutes into the flight, after you leave the atmosphere, the capsule windows offer free view of the surroundings. The problem is that you're sitting there well below the window level and the g-forces keep you down. You can't move, except for slightly turning your head. This is when a mirror on the sleeve of your launch suit comes very handy. By minimal manipulation of the position of your sleeve, I could find a reflection of the view behind me. The black of space and the blue of the Earth. As simple as that.

During my first flight, it was decided that we wouldn't remove the shades. It was thought that a visual reference to the external view might contribute to motion sickness. We had a number of valid reasons not to risk it. This time though, I was the only crew member who was on his second Soyuz flight and so I knew that I could do my job as an engineer and feel all right. I offered the crew to take this opportunity to look outside. I know Bob enjoyed it. Too bad Roman couldn't see anything from his central seat. Being a space vehicle commander comes at a price. I would trade it in any time though.

## ...Lena

That was a hoot! It was so spectacular! Even though, as I discovered later, I must have been not entirely conscious that day. There were a few things people asked me later (about something we apparently discussed that day) and I had no recollection of it whatsoever. Just don't try to tell me now that I borrowed money from you on the way to The Rocket. Because I was at another viewing site anyway.

God bless my friend Kathy Laurini. She's an embodiment of self-control, good humour and intelligent decisions. But most importantly, we can talk about all sorts of things no matter what current experiences we are sharing. This time I was picking her brain in my research for a story I wanted to write for a Russian magazine about American hypocrisy in dealing with gun control (or lack of it, if you ask me) and the resulting regular deadly shoot-outs.

Since the beginning of this year I was contemplating how to survive the launch and remain sane. I spent quite some time and effort in the course of the last year looking into various self-balance techniques. As a result of this, my iPod Nano was full of extremely useful meditation narratives that put me to sleep and enabled me to change my mind with relative effortlessness. I was sure that if I was in an urgent need of a distraction during the launch, because at some point it becomes all too much, the way to do it would be to close within myself with my favourite classic 'Daisy Pond' by Burt Goldman, who calls himself 'The American Monk'. Even though this nickname still sounds amusing to me, his life and work have impressed me to the extent that I refer to it in my occasional search for inner solutions and peace and recommend him to everyone (I never met him, he doesn't know I exist, but I'm honestly so impressed with significance and simplicity of his work).

On 7 May, I woke up with a deep conviction that Daisy Pond wouldn't work and I need something else for the launch. It had to be meaningful music. Too bad I don't know much beyond The Beatles. 'Don't stop me now' by Queen was topical, but has a slow start. And I prefer jazz anyway.

The traditional pre-launch behind-the-glass press conference was less hectic than promised, but more hectic than you would expect a press conference to be. Through the looking glass it was. As surreal as Alice's Trip to Wonderland. The wonderland of space was only a few hours away. Frank was looking out for the faces he recognized, and smiled and waved at all of them. I had a chance to give the pin and tie clip we made with the Expedition 21 logo to the ESA Director General. Frank was very pleased and thanked him for wearing it during their talk.

LEFT TO RIGHT: FRANK'S DAUGHTER NELE AND WIFE LENA AT THE PRE-LAUNCH CONFERENCE IN BAIKONUR

The following sequence is now a bit difficult to remember. Exit to the walk way to the bus, American astronaut Mike Baker (who was taking care of us) helping me to stand closer to the bus so that Frank and I could still have some fraction of an illusion of being together. Someone (of course) trying to push me aside. It helped to be able to speak Russian when you need to tell people not to touch you: they get the point faster. Frank in the bus with his hands on the window, me and Nele holding hands with him from outside. He blew a kiss to me and her. I thought it was jolly amusing to find later a photo of him blowing that kiss as a key feature on a page of a major Belgian newspaper. Yes, he did it in public view. But did those editors really think that it was a kiss to the nation?!

After a stopover for lunch and a short tour of the local museum, the bus brought us to the site where we were supposed to watch the launch. The Rocket was so small! It looked lonely but cheerful in the middle of the steppe. The steppe can be an awkward place to visit. It was hot and bright and perfectly windless. Everyone had dressed as lightly as they could (my deepest sympathy to the top management in their full business attire). Open shoes were a daring compromise for surviving the high temperatures. Strange insects on the ground. A peculiar bright green spider-shaped creature jumped out of nowhere on to Nele's leg. My flag with the mission logos came handy, we used it to shoo him away. We are supposedly more intelligent than spiders. But I'm sure you would agree it has happened to most of us – being hurt by something you perceived as inferior to yourself.

Tom, Nele and Koen were wisely spending some of their time in the shade of a strange little building – the only one for miles around. It looked like an electricity transformer box. The only other manmade construction was a wooden terrace where you could hide from the weather while waiting for the launch. But it had a roof that would obstruct the view. Francois, Frank's doctor during his first flight, was with us and had managed to find the one chair in the entire desert for Franks' mother Jeanne. She and Carine (Frank's sister) also preferred the steppe over the shaded terrace.

"Focus on The Rocket and send it good energy," I thought to myself as I walked out of the bus (Kathy was with a different group, we only met later). It would be useful for the situation and will give me something to do. I even tried to play Burt. It did not work. The general noise surrounding me made it impossible to hear the words. The Rocket was clearly planning to derive all the energy it wanted from the tonnes of propellant and intelligence of the people who built it. Ouch, my long-conceived masterplan won't work. Go for Plan B.

The situation kept becoming more unreal. Watching a launch is an exciting fun thing to do! Since Frank and I met, we like to do each and every exciting fun thing together. This can't be right! Why is he not with me?! I'd already visited the space museum on my own and bought their entire supply of locally produced pins and magnets, all 10 of them (with the wrongly spelled name, De Vinne instead of De Winne), to give out to family and friends. It was time for things to get back to normal. But this was only the start.

About two weeks ago, when I was forced to stay behind in Moscow while the crew had to spend extra time before the launch in Baikonur, I had plenty of opportunities to ponder how to deal with the upcoming weeks and months. No matter how much time you spend in advance planning for something emotional to happen, it's always different when it's actually happening. I knew that. Two weeks before the launch, anticipation of the launch was entirely different than it was a few months ago and for sure entirely different to what I expected the whole month of May 2009 to be like a few months ago.

An obvious trivial but functional solution finally dawned on me. When you deal with a situation that is too big for you to absorb, the only way to handle it is to split it into smaller parts that are manageable. I started splitting Frank's entire flight into chunks. What could these be, I thought to myself as I helped my friend, artist Svetlana Pchelnikova, to finish assembling and spray-varnishing the dolls she had made on our request as souvenirs for Frank's mission. A baby-astronaut hugs planet Earth in his smiling quest to protect it. This doll was based on the logo of her own charity project 'Doll Parade for Children', whereby celebrities with the help of the doll artists produce collectable dolls that are later auctioned for the benefit of the children's heart hospital in Moscow (the original doll and a fantastic auction collection are at www.paradkukol.ru).

This resonated well with Frank's commitment as a free-will ambassador for UNICEF Belgium and he happily accepted this as a symbol of his flight. There were only one hundred of them produced. But trust me, they don't feel small when you have to pack and transport them!

"If he was flying for one week, would I be so shaken right now?" I wondered as I installed the boxes for packing the dolls on the floor. Probably not. I would just be worried about the expected aspect of the launch safety and everything else technical that comes with it. And then I would be worried about the landing safety. One to two weeks apart, every now and then, is such a regular occurrence in our normal life (what is normal is of course another matter) that it won't even be a problem. Eureka! From this moment on I declare to myself that this is a short flight. This gives me the right to limit my anxieties to the safety of implementation. For now, the rest of the mission doesn't exist in my mind. Once the launch and the docking are done, I will deal with three to four weeks at a time. He's been away for three weeks before. That was OK then, it will be OK now.

Frank and I had thought about a few fun things for him to do during the flight. I made sure that all the necessary small consumables were available in my personal parcels to him. Now I needed to work on the content. This will split the time into chunks that you can put behind you and look forward to the next one, for both of us.

With a huge sigh of relief, I pictured myself as Scarlet O'Hara who was the world master of postponing until tomorrow the things that upset her today. He will be gone with The Rocket, not with the wind, and he Will Be Back, even though he is not a Terminator. Life can be such fun if you just find the right angle to look at it!

Kathy and I were hanging out in the field. Fortunately for us, the spiders went for the younger women. Anticipation of stress is always a stress. Distract from the launch, go back to your interview about gun control or the lack thereof. We covered the subject but the launch was still not happening. A few minutes to go. Russians launch on the second. OK, jazz time, I put one earpiece into Kathy's ear, one into mine. Brubeck, Take Five, all-time favourite. If the timing in my launch playlist works out, we'll have a good switch over for the launch – I promised to her. The five- minute composition was coming to an end. No launch. I started replaying it again. A minute into my favourite twist by sax I 'saw' a vibration. I mean it when I say 'I saw'. It was at least a second before I heard it or felt it. I pressed fast-forward. Kathy looked at me and exploded with laughter. The sound of the mega-roar and vibration was coming out of my iPod Nano. They say a picture says a thousand words. At this point, a soundtrack would have done just as well. I hope you know the tune in your mind when you read the lyrics. It was a hit in the late 1980s, and since Frank is such a big believer in a strong united Europe, this was the extra quantum of information from the Universe that I needed to finally figure out by the evening of 7 May which song I had to buy for 99 cents on iTunes to make the launch successful.

FRANK DE WINNE AND CHRISTER FUGLESANG WITH A SPACE GNOME GIVEN TO CHRISTER BY HIS WIFE LISA WHEN HE STARTED HIS SPACE TRAINING IN THE EARLY NINETIES AND A BABY-ASTRONAUT PROTECTING THE EARTH MADE FOR FRANK AND LENA BY SVETLANA PCHELNIKOVA

*We're leaving together*
*But still it's farewell*
*And maybe we'll come back*
*To earth, who can tell?*
*I guess there is no one to blame*
*We're leaving ground*
*Will things ever be the same again?*

*It's the final countdown...*

*We're heading for Venus and still we stand tall*
*'Cause maybe they've seen us and welcome us all*
*With so many light years to go and things to be found*
*I'm sure that we'll all miss her so.*

*It's the final countdown...*
*(Europe/J. Tempest)*

We danced to this song. I waved the flag with the mission logos. Our own soundtrack playing over the external reality turned The Launch into a private movie that I watched in the company of a very good friend. We laughed because it turned out to be the most unexpected twist in our experience. And simply because we were happy. There was a perfect blue sky, as good a visibility as you could possibly hope for, a flawless launch, kids and mothers and sister were feeling fine. It was a hoot. And it was spectacular. I quite easily convinced a few friends to stop crying. That was actually rather ironic. What else can you possibly ask for?!

\* \* \*

'Krainii' Airport in Baikonur is a small building without passenger facilities. Unfortunately it's strictly forbidden to take pictures there. I keep wondering if it's because of some secrets invisible to the naked eye or because it looks embarrassingly improbable that a facility like that could be a reliable component of a very complex operation.

By the time we drove to the airport more than an hour passed from the launch and much longer from the time we were last near any kind of civilisation. Of course we were a few women now in need of a bathroom. Making a loud proclamation of being a wife of an astronaut who just has launched into space gets you a few, albeit reluctant, favours in that environment. With the support of NASA staff, who helped me and our family in every way possible throughout this surreal trip, I managed to convince the local police to allow a bunch of us to use their men's room. This was a small victory of the human condition over dictatorial prescription of military conduct.

For the first time in my life I entered a plane at the rear. When we arrived three days ago we just walked down the stairs as you would when you disembark any cityhopper plane. This time the plane was preparing for departure while we were still boarding it. The engines were running. The noise was unbelievable. People were keeping their fingers in their ears to bring the din down to a tolerable level. Have you seen the scene in the movie 'Cabaret', where the character of Liza Minelli drags Michael York under a bridge as a train passes above their heads. "Aaaaaaaahhhh!" she shouts at the top of her voice. He looks at her in astonishment. And a second later, he also shouts with wide open mouth, "Aaaaaaaaaahhh!"

"Koen, shall we do some anti-stress?"

Koen is an angel. He is always up for games, fun and open experimentation. About three years ago, he organized for a group of his friends to shoot a movie called 'Back to Mars'. I featured in it as a part of the Mission Control team. Koen played one of the main characters who was an ex-criminal, but who'd been recruited for a rescue mission. He dies on Mars in the last scene of the movie. Speaking in their language, I was ROTF LOL when I saw the scenes in which I hadn't participated – he'd managed to get himself arrested in order to play his criminal role. I have probably told him that before. But what I've never told him is that I actually cried when I saw the close-up of him dying on Mars. I know, he's family and all that, but still, he was really good! Three years and a lot of movie work later, Koen together with his friend Michiel Knops are working on writing their own screenplays and are looking to become professionals in the film industry. Their current work is at *www.geheimnummer.be*

"Aaaaaaaahhh!" we shouted at each other as loud as we could without being able to hear anything. Anti-stress rules! High five! AAAAAAAAAAHHH! I think we surprised a couple of people waiting outside with us. I'm sure they forgave us (even though I won't be so sure they understood). We didn't disturb them anyway. They were keeping their ears plugged to survive the noise.

# 3 The docking according to....

## ...Bob

The only Station I saw when we were approaching was on the diagrams on the screens of our Soyuz vehicle. Two days in the Soyuz were exciting but limiting. Physically limiting. The capsule itself and the 'habitability compartment' is all you have. For the capsule, imagine three of you sitting tight next to each other in more or less an embryonic position. You assume this position as you settle into your seat a couple of hours before the launch. The seats are individually moulded to match your own body, so it's very comfortable, but it's very static. The seat liners remain yours for the duration of the mission. If you go back with another Soyuz vehicle, you take your seat liner with you and insert it into that vehicle. The moulding is done during a dedicated session by lowering you into a gypsum tub. You go in and out on a hoisting device to make sure that the form remains perfectly matched to your unique body shape.

The habitability compartment is the same diameter as the capsule but it's a cylinder rather than a sphere. They fit together under the fairing on the launch vehicle, and eventually this is all you are left with after you enter orbit. For the trip back, only the capsule with three people in it returns. It takes more or less 48 hours to reach The Station. We had to sleep twice during the trip. We just tried to stretch in various positions and not to bump into each other. Due to the weightlessness, you don't need much additional comfort. Due to the limited space, you don't need many fixings, just mind not to hit each other or the control panels by accident. In the Soyuz capsule, there are various controls and monitoring screens and buttons above the seats. To reach them you use a special stick (the pointer, or 'ukazka' in Russian). Everything around you is either fitted with more panels or with cargo bags. In short, Soyuz is physically limiting.

And then we touched The Station. It came out of the deep sky blue into my window and it was just magnificent. It was a surreal experience that made my spaceflight ultimately real.

We never discussed in advance in which sequence we would enter The Station. I was sure that Roman would go first. And we found ourselves having the funniest dialogue while the hatch was opening, which literally went like:

*"You go first."*
*"No, you go first."*
*"You're the commander."*
*"Well, that's why I go last."*
*"Then you go first."*
*"No, you go first,"* or something along those lines.

LEFT TO RIGHT: FRANK'S SISTER CARINE AND SON TOM AT THE DOCKING

LEFT TO RIGHT: FRANK'S MOTHER JEANNE AND BOB'S MOTHER EVA AFTER THE DOCKING

Roman insisted that, as commander, he was the last to leave his vehicle. Frank insisted that he has been to The Station before, so I had to go. This was the most unexpected and graceful gift from my crew, to have the honour to be the first to cross this symbolic divide. Mission Control told me later that it was a great relief for everyone on the ground to see us arrive safely.

People still keep telling me that I had the biggest smile on my face as I floated into The Station. It was the greatest privilege in my life. My crew spontaneously insisted on this. Lena said that the sequence of this section in the docking story will be the in the same sequence as we entered The Station. Well, this makes sense.

## ...Frank

It's a great responsibility to be a commander of a Space Vehicle. 'Great' as in 'big', and also as in 'fantastic'. I still hope that one day Europe will have its own manned spacecraft. Most probably I'll be too old by then. But still it's nice to dream. I enjoy having a serious responsibility and knowing what I'm doing.

I had to sit back and observe Roman commanding a flawless docking. I couldn't see much from my seat. He had a hard job, the visibility came too late. He's so natural at it. It must be in his genes.

BOB THIRSK IS THE FIRST ONE TO ENTER THE INTERNATIONAL SPACE STATION

It's the second time that I arrive at The Station. I won't say that it's a return home. Not at all. Home is on the ground where I live together with Lena. This is more like a very exotic hiking trip. You don't think you will make it to such an unusual destination twice, but when you do it gives you a feeling of unbelievable connection and familiarity.

Gennadi, Koichi, Mike, it's great to see you! We are finally there.

Lena told me yesterday the story about the song, The Final Countdown. Well, I'm in the final countdown now for about 190 days, starting from the departure from Star City to Baikonur. It's up to me to make this half a year a great experience. I know I will. I know Lena will help me a lot with it.

## ...Roman

After 11 years of training, five trips to watch others launch from Baikonur, working in Houston and living separated from my family, it has finally happened. I'm not sure if I'm awake or if it's a dream too good to be true. In the past 11 years, when it seemed that my flight was a given, several times some unexpected things happened and other people got flight assignments ahead of me. Until we left for Baikonur, I still couldn't believe that it would work out this time. I'm still getting used to the idea that I'm finally here, that I have already accomplished 50% of my task as the Soyuz commander and that we're so lucky to be the first on board with a six-person crew, representing all the five partners. And we all happen to be friends and get on well as private people.

CLOCKWISE FROM THE TOP: ROMAN ROMANENKO, FRANK DE WINNE, MICHAEL BARRATT, GENNADI PADALKE, TIM KOPRA, BOB THIRSK

The final approach to The Station was heavy for the head. We were under the impression that something might not be going quite right with The Vehicle. This is the trick that weightlessness plays on you. Since you don't have the force of 1g as a reference for your body, any force applied to you feels heavy. Even though the telemetry was fine, it felt like the capsule was shaking and swinging left and right, a bit like a plane in turbulence. I was expecting any moment an instruction from the ground to take over and switch to manual control. There were also moments when this could have been my own decision. Only later when I watched the footage of our approach filmed by the The Station crew I realized that everything was indeed fine.

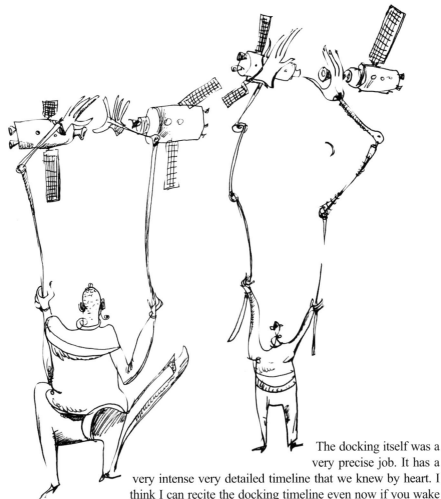

The docking itself was a very precise job. It has a very intense very detailed timeline that we knew by heart. I think I can recite the docking timeline even now if you wake me up during the night. I know now why I didn't study much poetry in school: I saved all my memory capacity for learning the procedures that make spaceflight possible! Up to the first touch between The Soyuz and The Station, each moment is prescribed. We implemented everything meticulously, monitored thoroughly and followed up precisely. Even though it was the middle of the day in Europe, we docked during the night, in regard to time on The Station. The Station makes a full orbit around the Earth in about one and a half hours. Half of it is a day and another half is a night. I know it was a day time docking for you. It was great because you could follow it up and have a party in Star City afterwards. But for us it was a night docking. It was only 15 metres in front of The Station until I could actually see the docking target – the so-called 'cross'. The ground was not hearing us. It was entirely my responsibility to make it work.

The real 'wow' moment for me was when we were docked but before we started opening the hatches. For the first time I had a chance to look outside through the 'blister' (the observation window of the habitable compartment of Soyuz). This is something that can't be trained for. I saw through the window the American segment coming out towards me. It was white and huge and clean and huge and real and so pristine and huge and well-lit and simply stunningly beautiful! The only time in my life that I had this feeling was when I stood for the first time at the foot of a skyscraper. I so badly wanted to go up the very top floor of that skyscraper but it wasn't possible. I was so badly looking forward to entering The Station. I was about one hour away from it and I knew that finally it would happen.

I'm not sure if this is a Russian thing or universal, that the commander should always be the last to leave His Vessel. It was imperative for me as the Soyuz commander that I honoured the moral commitments to my crewmates as much as I do my job correctly. They had to enter The Station before me.

## ...Brenda

Bob left home at the beginning of April for his pre-launch Star City session. With all the house chores to run and the Atlantic Ocean in the middle, there was no way for me to travel in the period before the launch to spend any time with him. I know a lot of people in the space industry ironically refer to making a cross-Atlantic trip as 'hopping across the pond'. But trust me, it only sounds easy. That is why the three days in Baikonur were even more so a painful treasure that we could share and cherish before yet another extremely lengthy separation. Treasure – for all the obvious reasons. If I honestly face this and the preceding few years of training, I couldn't tell you how many cumulative years we have spent apart. There is a different kind of bond that develops between the people who love each other in the absence of physical contact, however, that I equate to what happens when one's sense is compromised or lost. Painful. It wouldn't be easy to reactivate this process after the three days we cherished in Baikonur.

Baikonur. Oh well. Sometimes I was glad I couldn't understand the Russian language. I was just doing what I was told was OK to do. But some of it was real strange to me. From what I could sense, there was a lot of tension surrounding our presence. I still wonder why: we were there to embrace and support our husbands before half a year separation. This is all there was to it. Having our husbands off the planet on a totally different schedule and with different pressures would increase so much the sense of loneliness. Later into this mission, I also found that the lack of privacy put us in a pressure cooker. No one really understands that.

Sometimes two days can be a very long time. How shall I put it? Two days of waiting for Bob, Frank and Roman to reach The Station felt very different to a two-day spa retreat we took once a couple of years ago. But I knew it must have been rather different for the guys. I almost envied the uniqueness of their experience, their solidarity and bond. You could feel how connected they were, how much they relied on each other and trusted each other. The capsule was so small, and their bond was so big! And at the same time I knew that the connection and solidarity I had with Julia and Lena was also great. We didn't have to spend any time together and we didn't have many reasons to be in touch. As a matter of fact, Julia and I don't even have a common language, and yet it was a blessed gift that we could embrace each other as we shared the expectation of the launch and the docking.

It was with an impossible mix of excitement and relief that I saw my husband floating into The Station. Roman and Frank gave Bob the honour and pleasure to be the first to enter The Station. The energy inside Mission Control rose with the appearance of all three crew members as they joined Gennadi, Mike and Koichi already on The Station. This image still stands out for me: Bob coming through the hatch. Bob smiling as the happiest person in the world, which he truly was at that moment. I knew he was grateful for the crew decision to offer him the opportunity to enter first.

I was also immensely grateful for being able to witness history in the making. There were so many 'firsts' in this mission. Our guys were joining The Station to become the first permanent six-person crew, the first time all the five partners were on board at the same time, the first Canadian long-duration mission and the first European commander of The Station. It was truly a privilege to be present at the Russian Mission Control, a place where years ago, I don't think this would have been possible.

It's a done thing to arrive in Mission Control a couple of hours before the hatch opening. First we had an opportunity to witness the first contact, the retraction of the Soyuz vehicle to The Station and the complete docking. Only 90 minutes later did the hatch open. On one hand, it was a very long 90 minutes, but on the other this gave us time to mingle with the people. It was a precious opportunity to see a lot of familiar faces all at once, gathering around the moment that signified a dream come true for my husband and his friends.

After a greeting by the top management of the programme, a brief opportunity to exchange a few words with our spouses marked the beginning of a new segment of our lives. I figured some controversy surrounded the fact that we got a chance to talk. Yet again I was happy that I didn't have to understand what the problem was. I just enjoyed this little gift. More clearly than ever before, I faced my new reality: this physical separation left us with the task of communicating via intermediaries or facilities in the realm of reduced privacy.

But there was a party to host in Star City and the mood here on Earth matched that on the ISS. From my personal home Mission Control on the 'other side of the pond' I could only be virtually involved in organization. Traditionally, it's the Russian wives who host docking parties. Lena, being Russian by birth but Dutch all her grown-up life, together with Julia, made it a joint spouse crew event where I was happy to join. This was our contribution to the glorious list of 'firsts'. I was happy to be a co-host of what I understood was the first joint international docking party hosted by the crew spouses. It felt so natural for us to be together, the three of us, sharing this powerful but transient moment in our lives.

# ...Lena

In the middle of March, during the closing training session week in Cologne, I had caught a really bad cold. It was one of those misleading spring weeks with warm days and freezing nights. It was entirely my fault, because I didn't bother to dress warmly when going between the guest house and the main building of the Hotel Zur Quelle; where visitors coming to EAC often stay. This was also when I properly met Julia. We met once briefly the previous December at a big party but we didn't have a chance to talk. The weather was good, Julia was great, but my cold was terrible! The nose-related symptoms were gone as prescribed within seven days of treating them, and one week of not treating them, but the cough wouldn't stop. It faded out for a day or so on a couple of occasions, but then came back. It was exhausting and it was humiliating: I couldn't have a decent conversation. I knew that any sentence longer than three words would be interrupted by bouts of unstoppable coughing and I had to keep saying 'excuse me' every few sentences. I tried various medications but it wouldn't help.

I was wondering how annoying people must be finding this, talking to me as I kept coughing and apologizing. Curiously, I didn't cough when I didn't talk, only at the end of a day if I spent the day in meetings and discussions. At some point, I realized that the intensity of my cough depended on where I was and with whom. I couldn't help but recognize that I burst out coughing even before opening my mouth when I had to deal with subjects or people I didn't want to face, but couldn't avoid. Am I really having for the first time a clearly pronounced psychosomatic condition? This can't be right. I've had long coughs before. And yet, on private weekends when we could go away from Star City and leave everything that had to do with Frank's work, the cough would drop significantly. At some point I stopped resisting it.

As long as it didn't keep me awake at nights, I was OK, I kept repeating to myself. The only other way to take it would be as a message from the Universe to talk less. And that wasn't fun. First of all, simply because I'm a woman and, in line with every joke, I'm entitled to talk a lot. It's really important for me to spend time on the phone with my friends. Star City is a nice quiet place. But it's too far out to be able to see anyone who doesn't live there. I needed my regular 'support circle' to remain grounded.

Do not dare to cough in front of the Russian quarantine doctor who takes your temperature and checks your throat when you get your chance to see Frank in the last days before the launch at Baikonur. I kept convincing myself.

Frank's doctor agreed. There was nothing wrong with my throat according to the good doctor.

It would really be important for me to stop coughing, I was thinking on the way back from the launch. I wanted to make a speech, thank everybody who helped us, give out a few souvenirs at the docking party that I was hosting together with Brenda and Julia after the docking. I couldn't do it if I couldn't be sure that I won't cough in the middle of it.

Julia was taught to drive by Roman. It sounds like a dream, doesn't it, a few driving lessons from a jet pilot! The misleading appearance of a soft blonde totally clashes with her ability to overtake aggressively in a side lane. If you have ever driven in the suburbs of Moscow, you would know that the sides of the roads gradually merge into compressed soil that gradually merges into mud. All the way from the Mission Control, she was speeding and overtaking one solid traffic jam after another by driving on the soil on the edge of the road. There was a party to host. Being late would not only be embarrassing, but also highly disrespectful towards the community. Only thanks to Julia's driving skills and focus did we make it.

I think it's the first time ever that all three spouses (only one being of the Russian crew member) joined forces and united the docking reception under one roof. It was a really nice crowd from all over the world. Our families and friends, who had travelled to the launch, and people who had worked with us and our husbands in the last few years.

I am so lucky with the girls! This time Julia and Brenda supported my creative deliberations and agreed that we make an invitation to the party that would look different to the typical Star City invitation. A few weeks back, Julia and I had a chance to join the Star City tour with the Belgian press. We didn't think then that this picture of us taken in our husband's Soyuz simulator would turn out, not only as a special cheerful private memento for us, but also as the most obvious feature for the docking party invitation.

Communication for The Station is provided via various technical means. One of the options is to use ground stations on the territory of Russia. Each ground station has a circular coverage area. They are located in such a way that they provide pretty much coverage over the entire territory of the country. Typically, the launches of the Soyuz vehicles and their dockings are done with the help of communications provided by the Russian ground stations.

The technical steps related to opening the hatches take some time after the initial docking of the Soyuz to The Station. One and a half hours later, or one orbit, the hatch opening takes place. Similar to the launch, this is a festive occasion that brings together not only friends and families and the people who work to make it happen, but also the highest management of the participating agencies. The hatch opening and entrance of the new crew into The Station signifies the second successful milestone of any mission to The Station.

Even though the docking party is a 'spouse thing', Frank and I thought it would be neat that the crew themselves announced the party while we – the wives – just hosted it in their honour and on their behalf. We had to take into account that this is a big, but nevertheless private party. How could we make the invitation in such a way that it was a nice touch for those who are invited but does not draw the attention of anyone else? We made a printout of the invitation card and Frank included this as one extra page in his Soyuz onboard documentation, with the intention to show it briefly during the hatch opening video conference on entering The Station.

To be honest I didn't think that this would work out. It was a nice enough idea to try it but it wouldn't have been a big deal if it didn't work. Often when you positively embrace things but don't get too eager about them, they work! With astonishment I watched as my husband pushed a piece of paper floating in front of him when he entered The Station and turned to the screen. It wasn't very visible, but those who knew the subject would recognize it anyway. I'm sure that on Frank's return this picture will take its well-deserved place in the space memorabilia collection that occupies a whole cupboard in our spare bedroom.

The conference at hatch opening is held by the top management of the partner agencies. Time permitting, spouses are usually offered a few words to say to their husbands. Chris Hadfield, part of the back-up crew for this mission, was hanging out with Brenda and her family and of course was including us into his scope of attention.

"Hey, Lena, when you talk to them, you can tell them that the back-up crew is still ready to fly if they've changed their minds," said Chris.

"I sure will!"

I was mentally rehearsing something short and witty to say. This suggestion was perfectly set up for my category of witty. I was just terrified that I would burst out coughing, which I still couldn't control. There is the famous 'KISS' principle in business management. *Keep It Simple, Stupid.* Simple and Short should be my personal version of KISS on this occasion, I thought to myself. Brenda wanted me to go first. Julia was hesitating whether to talk at all. I was

DOCKING PARTY INVITATION

wondering how I could make a joke based on the famous 'White Sun of the Desert' movie that all the crews watch the day before their launch. The plot of this film involves a bunch of women who are rescued from a harem by a Red Army officer. Following military discipline, he tries to appoint one of them to be in charge of the group. They misunderstand and think that she has been appointed as his favourite wife. I was thinking how to make a short and obvious joke about having been appointed 'the opening speaker wife'.

The conference was coming to an end. Brenda, Frank's mom and I were hanging around behind the seats of the management who were talking to the crew. This is where Joel Montalbano, the Head of the NASA office in Moscow, suggested that we should wait to make our short greetings. For obvious reasons, Brenda, Julia and I try to do things together that apply equally to all of us. But we are taken care of by different systems. Brenda and I being the spouses of the Western astronauts get support from NASA, Julia gets taken care of by the Russian side. If Bob or Frank had flown on a contract with the Russians, rather than as a part of their own agency contribution to the programme, we also would have been taken care of by the Russians. Unfortunately it felt, on more than one occasion during the launch and the docking, like the formal aspects of this situation were far more important to the Russian management than the fact that they were dealing with people who are undergoing a very intense personal experience.

In the English language, 'human spaceflight' used to be called 'manned spaceflight' until a few years ago. Then political correctness that had to be respected on various levels of American and European society made it necessary for some to change the term. I didn't get far with my jokes that if you really want to be fair and equal, start by calling it 'womanned spaceflight' for the next thirty years and then switch to 'human'. Another suggestion would have been to be consistent and change the term 'unmanned' (for cargo vehicles or other satellite launches) into 'unhuman'. Jokes aside, I think the West has not only changed the name, but also further refined its attitude in general to the human aspects of human spaceflight.

Russia, on the other hand, didn't have a reason to change anything. The Russian term for human spaceflight is literally 'piloted spaceflight' (as opposed to 'unpiloted' for unmanned) and correctly represents the meaning of the activity. While I've met a lot of nice, kind and open people on the Russian side who were looking at various ways within their individual responsibilities to help us and make this experience lighter and friendlier, the Russian system seems to reject the very notion that it is working with humans – the crew and those who are close and dear to them. It felt like we were treated as highly trained achievers for the flight and as people who unfortunately needed to be taken into account. Far too often I felt handled as an item, rather than supported as a person. The saddest thing is that it takes only two or three people at management-level to be formalistic, bureaucratic and unkind to completely destroy you in this situation, when you are already highly vulnerable because of everything you are experiencing at that moment and have to face in the coming months.

Julia was hesitating if she should talk at all. The moment Joel told Brenda and I that we should be ready to talk, we went to ask Julia to join us. Joel was happy to support it. She didn't feel right about it, she should have been invited by the Russian party. She stayed in her seat.

A big boss from the Russian side announced that this call was closing, fully disregarding the fact that we were standing behind him with Joel suggesting to pick up another receiver to make it more convenient. At this point, another big boss from the Russian side pulled Julia out of her seat and forcefully stated, "You should let the wife of the Soyuz commander speak!" (emphasizing the words 'Soyuz commander'). He put the receiver into her hands. Julia said a few words. The Russian boss took the receiver out of her hands and said goodbye to the crew.

I was astounded that there were people in whose world being a Soyuz commander makes you more worthy as a human being in comparison to your crewmates (or to any human being for that matter). Forget me and Brenda, what about the crew on board? Aren't they supposed to contribute to a good atmosphere on The Station, not only in terms of oxygen balance but also in terms of moral comfort? Don't these people understand that intentionally denying a ten-second hello to Bob and Frank makes it not only unpleasant for them, but also puts Roman into an awkward position that he gets unfair advantages that are not related to differentiation to their professional tasks and therefore simply should not happen?!

At this point, Joel picked up another receiver in order to save the line and put it into my hand: "Go ahead, just speak as we discussed."

Thank you, Joel. It made all the difference for all of us.

A few people suggested to me that I should not take so personally the aspects of the launch trip and the docking that were disturbing me so much. I don't agree with this. I agree that I don't need to let myself be emotionally damaged by the fact that other people lack comprehension or sensitivity in the situations where they would be most needed. After all, it's their karma, not mine. But I don't agree that something which is done to me as a person by people who know me is not personal. Rather the opposite, this is how these people choose, for whatever reason, to treat other people. I have no problem with them. I haven't done anything which gives me a bad conscience or keeps me awake at nights. But sometimes I want to look into the eyes of those who, with their behaviour and attitude, are telling me that my husband's desire to have me around supporting him and my desire to be there for him are wrong. I want them to tell this to my face. Why wasn't I surprised that those few people were avoiding eye contact lately? I move on in my life in the blessed company of many enlightened people who come my way. If those few who choose bare formality over human positivity see themselves good for treating unkindly the people who are in a weaker position and depend on their values and attitudes, it remains their choice. I wish them well on their path.

GUY LALIBERTÉ

I was too much taken aback by how the things evolved in the last few minutes and limited my speech to the expected short 'we're fine and we love you' to make sure that there is still enough time for Brenda and Frank's mom to talk. I completely forgot to give the message from Chris that he was ready to fly. Chris was cool. He forgave me immediately.

Another sweet touch that Frank and I contemplated for the docking party was that they would join us by a phone call from The Station. Julia arranged for a microphone in the reception room. The idea was to put the phone to the microphone so that the crew could welcome the guests and thank them for everything they have done for the mission. Unfortunately some hissing and crackling between the phone and the microphone made it almost impossible to hear what they were saying. But nevertheless we succeeded in having our crew in the party dedicated to their successful docking.

It was one of those good parties with a lot of people you are happy to see around. Formal speeches, which took place as expected at the start rather quickly reduced to informal toasts at each table. The standing reception has a huge advantage over a sit-down dinner – it allows you to communicate with everyone in the room.

The new spaceflight participant Guy Laliberté had just arrived at Star City in the middle of May. We also invited him to join. A bunch of people from the Belgian embassy came to Star City because they felt that it was important for them to be part of this event. Unfortunately Moscow traffic on Fridays is so bad that they only made it for eleven that evening. They were extremely happy nevertheless.

The party started slowly fading out after midnight. Kathy had to leave for Moscow to catch a plane to Houston early the next morning. Unexpectedly it was the first sign of the end of an era for me, an era of preparing for Frank's spaceflight. I got a lump in my throat as I hugged her goodbye, in expectation of having to go back to virtual contact for many months to come. On the walk back from the Dom Kosmonavtov building, Doug, the Canadian crew surgeon, was telling me that the next time I talk to Frank I absolutely had to thank him and Roman on behalf of the whole nation for offering to let Bob enter The Station first. I knew what he meant. We all gasped and applauded to Bob's smile. I've known Bob as a positive cheerful man for many years, but I've never before seen his face in what they call a smile from ear to ear! Brenda told me the other day that this is now The Picture of Bob she has on the fridge to say hello to every morning.

We went into Shep's bar in Cottage 3 where the party naturally turned into a regular weekend evening for the ground inhabitants of the Western segment of The Station. Chris Hadfield was playing guitar again and singing, patiently encouraging everyone else to give it a try (I never understood why he didn't become a musical pro). The newly arrived team for the next space flight participant Guy was getting to know us, finding their way in the social part of the Russian space training world. Frank's son Koen was getting pool lessons from Helene Hadfield. She turned out to be the best coach ever. Koen featured in a video I recorded for Frank later that night with a clear promise to always beat him at pool from now on. This was another one of those extended family evenings we had every now and then over the last four years when we lived on and off

in Star City. This is one of the things that I will miss the most about the previous four years of my life.

I never made that speech at the party. The party took its own course. The course was nice. There was no point in creating an extra wave. I realized the morning after the docking that I was not coughing anymore. I haven't coughed since. Maybe just once or twice while writing this chapter.

## ...Julia

We were running seriously late for the party. Docking on Friday evening on one hand was a great comfort because it naturally puts people into the weekend party mood. But on the other hand, it made the travel between the Mission Control and Star City after the hatch-opening in the late afternoon a complete and utter nightmare. The way out of Korolev, where the Mission Control is located, in the direction of Star City goes via a so-called 'drunk road'. We're not sure where it gets that name from. Maybe because it's bendy and you don't want to attempt navigating it when you're drunk, or alternatively because it's narrow and hidden and you're safe not to meet any police there, so it's a safer drive when you're drunk.

In any case, this single-lane country road is a famous connection between two neighbouring highways. Of course it was at a standstill on the evening of the docking, Friday 29 May, around 7 p.m. This is the time when Lena, Brenda and I had asked the restaurant to open to start our docking reception. After the 'drunk road' reaches the town of Schelkovo there are several ways to continue from there to Star City. We were exchanging phone calls with cosmonaut Sergei Volkov who had left earlier and could tell us where he was or wasn't stuck. I took a small detour from the town of Schelkovo to the Schelkovskoye highway. It added kilometres to the drive but for sure saved us time. We were finally on a straight road. The day before Lena and I had spent a lot of time deciding what to wear for the party. This was our party, we were in all sorts of states because of the obvious worries and fears and stresses connected to the launch and the docking and six months of strange solitude, so it was more important than ever to allow ourselves to enjoy the pleasant little things like choosing nice outfits. I'm sure all the women would agree that it helps inner balance when you feel that you wrap correctly the outer surface.

We expected that we might be short of time in the evening. In order to save time after the docking on the way to the party, I stopped by Lena's room in the profilactorium in the morning and took her outfit for the night and brought it back to my apartment to hang it up together with mine. Up until the middle of our drive we were still discussing if we would manage to change and get going within five minutes. This would have added only an extra 10–12 minutes to our drive. By the time we were on the straight road it was clear that we were so unforgivably late that changing clothes was out of the question. Lena laughed and said something about the ironic sense of humour of her guardian angel, who was probably ruling her own views on life. "It only took me until two in the morning yesterday to decide what I will wear," said Lena. Lesson learned yet again – drop vanity. You're the only person in the world who cares how you look. Apart from those who don't like you and look out for your mistakes, but who cares about them. Oh well, all those little pleasures of being an allegedly mature woman.

As I was half-listening to the banter between Lena in the front passenger seat and Alexander Samokutyaev in the back, I saw in my peripheral vision something which didn't make any sense. It was bright daylight, around half past seven in the early evening. All of a sudden it was an unbelievably empty road. From my left side, a huge brown moose was slowly crossing the highway. I was driving fast. For a fraction of a second I hesitated, should I push the brake or the gas pedal. I chose the latter. We drove past him as he was approaching the line in the middle of the road. I've seen the road signs about crossing animals before, but as a matter of fact I've never seen crossing wild animals in my life (stray cats don't count)! We and the moose missed each other by seconds. He was like a cow but taller and completely deep brown. They say that a collision with a moose is as bad as colliding with a lamp post at high speed. I now show everybody the spot where we saw him. I'm not sure people believe me but this doesn't matter. What I believe is that such a narrow miss, the potential catastrophic dangers of which dawned on me only much later, was another good sign that everything will be just fine.

\* \* \*

The party is over. We made it. We ALL made it! The amount of time Roman and Frank and Lena and I spent on organizing the three events that are supposed to be taken care of privately by the crew and the wives is huge. The three events are: a celebration for the instructors on the completion of the final exams by the crew (up to 100 people can easily stop by), breakfast before the departure to Baikonur (anything around 50 people) and the docking party (another 100 people dinner event). We made it.

The docking party hosted by the wives is a tradition that comes from the days when there were no real organized parties but all the people in Star City would ring the doorbell anyway to congratulate the wives of the crew on the successful docking. Gradually this led to a tradition to host the wives' docking party in a place where service was offered and the wives had a chance to party together with the guests. There are two major locations for these parties in Star City: Dvorets Kultury (Palace of Culture – a typical name for a local venue in any Russian city) and Dom Kosmonavtov (House of Cosmonauts – another cultural venue specific to Star City). The latter hosts the Star City museum. This is also where the major official events are held such as 'vstrecha', the celebratory meeting after a landing.

We all decided that Dom Kosmonavtov was a nice place for a private party. Yanina and her colleagues run a great restaurant there, with a personal touch and caring attitude. We would recommend them to everybody. People in Star City, the spouses of the future crews are asking me nowadays if Dom Kosmonavtov is a new docking party location or an established tradition. I would be glad if we set this trend.

It was wonderful that Brenda could finally join us from Houston. I can only guess how hard this must be for her – an extra two months on her own before the launch, having to rely on external help to make her way around during these last few days, which are already very intense. The launch and the docking are done. We can finally breathe out, in order to take in a new deep breath.

# 2 Personal life during the flight, in space, on the ground

## There is none!

My personal life is circling Earth at a speed of twenty eight thousand kilometres per hour at an altitude of 400 kilometres, making 16 orbits per day, enjoying sixteen sunrises and sixteen sunsets in every 24 hours!

These physical extremes are hard to grasp in earthly terms, but they don't do anything to us apart from expand our comprehension of the 'near Universe'. When the Malaysians were training for their flight, one matter that came up for serious consideration was daily prayer. For Muslims, there are five prayers daily and they are connected to sunrise and sunset, and you're supposed to face Mecca when you are saying them. You have to wash your face before you pray and you have to kneel.

Two international theological conferences were held in order to determine the rules of correct conduct for the Malaysian astronaut who wanted to obey all the rituals correctly. A solution was found whereby he could face Earth in general and that would symbolize the right direction, he could wipe his face rather than wash it (this is as good as washing gets on The Station), he could say his prayers silently, and most importantly the days would be counted as on the ground, every 24 hours as opposed to The Station's sunset and sunrise rhythm. Otherwise he would have had to pray five times every one and a half hours.

Gosh, I wish all the matters of heart and soul could be resolved by a couple of conferences of people gathered in search of a positive solution. To make sure there was no ambiguity in the matter, a manual was issued 'A Guideline of Performing Ibadah (worship) at the International Space Station (ISS)', which was approved by Malaysia's National Fatwa Council.

My personal life in the days of the Malaysian flight training included occasional table-tennis matches against 'angkasawan' Sheikh Shukor in Shep's bar. I always lost. But I lost to the first astronaut of that country, so it didn't feel so bad. I didn't lose by much anyway. I thought this was much better for the Universe than if any astronaut lost to a girl!

The most fun I had in a table-tennis game was with Japanese astronaut Soichi Noguchi. We started playing and talking and naturally it turned into a Russian conversation lesson. Tennis racket in Russian is '*raketka*'. A rocket that flies into space is '*raketa*'.

'Raketka' also means 'little rocket'. You will see more about the Russian suffixes soon. One thing led to another and we slowly discussed the whole launch sequence and how not to confuse it with a game of table-tennis. Simple, unexpected, unforgettable and completely priceless.

Back to my personal life. Frank calls me whenever he has a chance. Communication coverage permitting, he calls first thing after his morning routine (wash, brush and planning conference). The Station typically lives on GMT. This is British winter time. This means two hours behind The Netherlands in summer. When Frank travelled to Houston for training and I stayed at home, I was completely wiped out by the end of his trip. Yes, I meant what I just said. I got up around eight in the morning my time and went to work. He would have called me on most days for a couple of minutes when he got up in Houston, around seven in the morning his time, two in the afternoon my time. I came home at some point in the evening, Frank's working day lasted until five or six in the afternoon, which is midnight or one in the morning my time. This was the first opportunity we had to talk on week days if we wanted to have more than a couple of minutes together. And we always did. Frank always told me to go to bed. I never could. He knew I wouldn't. He always called. I was finally going to bed around one thirty in the morning. I would get up around eight next morning to go to work.

Two weeks of this regime take a bit out of you. But I simply didn't see an alternative. In this most strange and exciting experience of sharing a life spread between the continents, the word 'sharing' always remained the key to everything. Frank is now in space and I'm at home. The phone line to space is open at this very moment. He called me a couple of minutes ago. And then the smoke alarm went off. When this happens everyone has to stop whatever they're doing and respond to this 'off-nominal situation'. He told me quickly: "Hold on, stay on the line." The loudspeaker stays open for me. I'm writing. I can't quite make out the words, but I can hear Gennadi in the background talking to Houston as I'm typing this.

Eventually the line drops out. Fortunately for my sanity we had spoken earlier today about safety versus functionality. The latest resistance exercise machine is currently broken. The crew are back to using the previous-generation machine, which is known to be somewhat less effective. The newer machine was flown up and installed only recently. It was in use by six people on a daily basis. It broke after one month. When I hear things like that I get terrified. You know why? Because in my linear mind, I think that if they can't figure out how to make some exercise machine work, how can they possibly be sure that The Station and its life-support systems work?!

There are zillions of exercise machines on the ground, there is a multibillion industry behind selling a promise of charm and longevity to the world through building trimmer abs and yet, developing something based on existing know-how for operation in space seems to be above and beyond the current level of any rocket science. It kind of bothered me. I asked Frank about this earlier today. He explained that there are components of the machine that can't be tested for long-term performance in space on Earth. This equipment was especially developed for space. But only now, after one month of operation, did it become clear which components wear out faster and why. The next Shuttle will bring spares. The crew is anyway trained to repair it.

"There has been a similar case with the computers on board, so why do you worry all of a sudden about an exercise machine?" mentioned Frank casually.

"What do you mean all the computers on board?" I gasped for air.

"Don't you remember, a few years back all the computers of the American segment crashed at the same time," Frank noted.

"How is that possible? Was there was no immediate evacuation?"

"No, because this was not an emergency."

"What? How can all the computers crashing not be an emergency?"

"Well, it is as far as the computer system itself is concerned, but it's not an immediate threat to crew survival," said Frank calmly.

"Is it because the computers on the Russian segment and on the American segment can take over control from each other?"

"That as well, but even if they all crash together at the same time, this won't be an emergency requiring an evacuation."

"Aren't the computers running the whole Station?"

"Yes, but it doesn't mean that for a period of time we can't control things manually while restoring them," said Frank matter of factly.

"But you have no contact to the ground?"

"Correct. But then we do have two Soyuz vehicles and they have their own MBRL (Russian abbreviation for inter-vehicle radio link). Emergencies that potentially require evacuation are fire and depressurization. The Station control computers have nothing to do with that. If necessary we can each be in our respective vehicles within two or three minutes."

As often happens in my life, I received something literally a few hours before I really needed it. That conversation earlier in the day, triggered by my concern over a broken piece of fitness equipment, gave me the peace of mind to continue writing right now. There is a procedure that if, God forbid, anything really happens, the spouses are notified immediately about the situation (a wise move – otherwise they would find out from the journalists ringing their doorbells at any time of the day). But Frank knew that he had to call back after the line held by my loudspeaker eventually dropped out. This kind of thing happens a lot anyway.

Indeed I heard from him about half an hour later. There was some overheating in the SRVK (Russian water regeneration system). They had to do some work to remedy the situation. Frank even wrote a few lines in his diary. He started by writing a private diary every day during the first two weeks of the mission, but then gradually stopped. Now he only opens it and puts in a few notes on the days when something particular happens. I think at the beginning it was a helpful way for him to set his own pace and structure, to ponder on his situation, to give a personal context to the new environment. Quickly we developed a routine that nicely matched his new habitat. We picked up our usual ways of spending time together at a distance, staying in touch and maintaining contact with friends. A private diary lost its therapeutic function and became a mere record of chronological sequences if they were worth memorizing. Exactly as we expected, this flight for us is a totally shared experience where the physical separation is just a mere inconvenience while we embrace and protect more than ever our connectedness and togetherness.

## Divine inspiration

I had a strange but nice dream the night I started writing The Book. I was attending a spiritual retreat where people came to learn to fly. Apparently flying was a matter of concentration of will and intention. It wasn't even flying. Flying is commonly associated with spreading limbs and waving. This was more like relocation through the air. Once you 'caught the muscle' of focusing your power on flying it came effortlessly, like everything you do when you possess well-developed skills for it. "How interesting," I thought to myself in my dream, "I wonder how long it would take me to learn this." I pushed off the ground and started slowly rising in a comfortable and effortless motion upwards.

VOLCANO ERUPTION WITNESSED BY THE EXPEDITION 20 CREW

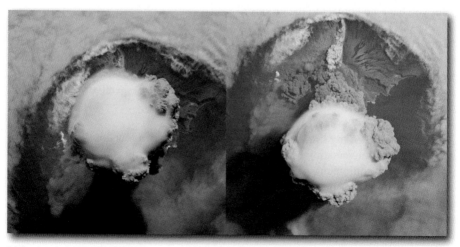

The landscape reminded me of the geyser valley where we went on holiday last summer with Frank and his son Koen. An endless horizon, pale orange light from the Sun, deep blue of the infinite sky. A few clouds were there merely to add a touch of reality to this heavenly picture. It was a place to breathe in freshness and to meditate. In my dream I made a very high but very slow jump to the height from which I could see the gathering of people attending the course. It was as if I was on the top of the mountain, with the only difference that I was actually in the air. Then I landed extremely slowly on the other side of the area where the group was standing. Probably my mind had short-circuited into a parabolic trajectory of the jump because the only way to experience 'weightlessness' on the ground is to fly in a plane on a parabolic trajectory (the way to do this for real is by going to the Zero-G company for such a plane ride). There were other people around me who also could jump-fly. Some others were trying in vain, being completely puzzled by the fact that it was at all possible.

Iceland to some extent was formed by volcanic eruptions. Most of the shoreline looks very unusual – like no other landscape you would find anywhere else in Europe. Some say NASA used to train its astronauts for the Moon landings in Iceland. I don't know if this is true but the road to the airport sure looked like it went through an alien dwelling.

Where did that come from? I woke up surprised. I reminded myself that, since I'm a psychologist, the secrets of the trade allow me in the best Freudian-Jungian traditions to interpret any dream in almost any way I wanted, as long as it appeared convincing and logical. Female logic will do. After all there is the classic story that one of the grandmasters had suggested to a patient to interpret his dream about a shoe sole that it was about his soul. Because it sounds the same and the patient's subconscious had, as it always does, masked its message in this symbol. Obvious, isn't it? Or it was a symbol of his subconscious resistance to learn spelling in the first grade at primary school. I decided to see my dream as a promise of a new opening in life and success on the selected path (I try to interpret all the dreams this way and strongly recommend this to everyone). Or did my brain just short-circuit to Iceland and the Moon and gave me 1/6 of the usual gravity for this jump? But a second later I realized that this must be inspired by the picture of Roman I received yesterday.

Frank sent me a bunch of pictures to distribute to our family and friends with an update about his flight. I mostly included pictures of him and views of the erupting volcano and rising Moon. Imagine the luck – flying over a volcano that only erupts once every hundred years or so, and there is good visibility! I think people who are generally lucky in those small but sweet things must have done something right in their life (or one of their previous lives if you believe in reincarnation, or in one of their projected future lives if you believe that time does not exist) to deserve such a harmonic support from the Universe.

And the Moon, yes, you do see it more or less every day. And yet, seeing something so familiar in such a new way – I couldn't stop staring at those pictures on my screen.

But Roman, one of the funniest and most cheerful men you could ever meet, for as long as I've known him he always is the embodiment of good-heartedness, humbleness and never-ending humour. After 11 hard years in training and a few unexpected situations that delayed and derailed his launch assignments, at one month into the flight he still seemed to enjoy himself to the extent of being euphoric. It was such a delight to get a glimpse of him every time. His usual comic self at home in an environment that allowed the pulling of more pranks.

It reminded me of Frank's stories of how solo pilots play with the clouds – going up and down, inside and around. I never quite understood if it was allowed, or if they did it in a secret aspect of their real work. Squadron life in the air force is something that deserves its own book. Only towards the end of the mission did Roman say that he maintained this sense of euphoria on purpose: it was the easiest way to deal with everything you miss too much. Space is work. Life is at home.

But back to Roman, I think he had meant to impersonate a character from The Matrix in this photo. I love comedy but I rarely laugh out loud. I exploded with laughter when I saw this one. It was too good to be real!

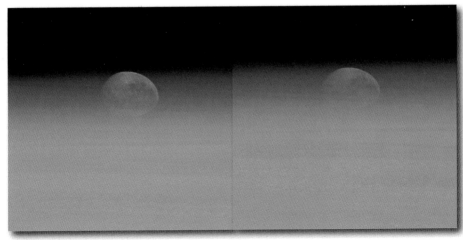

ROMAN "IN THE MATRIX"

I didn't jump into The Matrix movie in my dream. I just softly went up and down across a geyser valley full of people attending a spiritual retreat. Maybe the book 'The Moon is a Harsh Mistress' by Robert Heinlein got pulled out of my subconsciousness and connected into this dream too? In that book, Earth people take time to adjust to the environment on the Moon. Mind you, they don't have fresh air and they create a revolution under the guidance of a big computer that had developed a soul. I read it a few years back because Charles Simonyi took it to the ISS during his first mission, along with Goethe's Faust. Those two books for him represented a literature phenomenon that transcended cultures and united humanity.

## Like Captain Jean-Luc Picard

As a young boy, Frank was fascinated by the 'Star Trek' series. Captain Kirk, with his interplanetary crew, was boldly spreading ideas of peace and humanity, taking unforeseeable risks, getting things to work out fine for his starship and his galaxy, and never compromising on the human touch and human values. He was a real hero for Frank. Kirk stood up against injustice, and most of all he believed in the infinite strength of the human spirit and exemplified persistence in the tireless pursuit of boundless exploration.

Chekov: "Course heading, Captain?"

Kirk: "Second star to the right and straight on 'til morning."
*(from Star Trek: The Undiscovered Country)*

The idea of sitting in the commander's chair (which looked rather like a futuristic throne), overlooking those big monitors with sketchy outlines of the few neighbouring galaxies, ordering the use of the faster-than-light warp drive to arrive at the next destination, and having to cover distances of such magnitude that orientation towards one star was precise enough, gave Frank a spin for flying.

He believes more so now than ever in human exploration, which is inherent in human nature and spirit. Unlike in some Star Trek voyages, and unlike in relatively recent Earth history, we are fortunate to live in a day and age where exploration doesn't need to be accompanied by potential invasion and conquest. We are living in the third millennium where all the exploration we do is for peaceful purposes and for all humanity. Space exploration transcends cultures and brings benefits to everyone on Earth, no matter what their origin or beliefs.

Some of our findings today might only take on their full meaning and open their full potential some decades in the future. After all, how many Europeans in the first half of the 16th century benefited from the fact that there was a continent on the other side of the Atlantic Ocean accidentally discovered by Columbus? What we are doing today in space exploration is greater than us. We owe it to the future; we owe it to the life on our planet Earth. We owe it to ourselves, to reach out further and to advance the legacy of the human spirit and human exploration.

Space exploration is one, albeit the most obvious, way to explore. Frank is an explorer by his inquisitive nature. He explored the sky as a military pilot. He is exploring higher levels of the sky as an astronaut. He applied for the astronaut selection when everyone told him it was a lottery that was impossible to win, but he won. He believes that apart from his qualifications that allowed him to pass, there was a lot of good fortune in his circumstances. There are definitely some tens of thousands of people in Europe who are now fit and sufficiently well educated to learn the job of an astronaut and do it very well. At this moment there are just not enough jobs. He is doing his best to take forward the boundaries of the space exploration, in his personal quest for more people on Earth to have an opportunity to become part of it.

Not only did Frank become a European astronaut. On this flight, his second, he became the first European commander of the International Space Station. It has become tradition in NASA to produce unofficial crew posters with the themes of famous movies or television programmes. As always, the commander has the privilege to make suggestions to the crew for the theme to use for their poster. As for a number of other fun things on Frank's mission, I also had the privilege to contribute ideas. We started with Belgian history and literature. The most famous Belgian astronauts of all, Tin-Tin and his dog, were a crew of two. No matter how hard we tried, there was no way to play on this image and arrive at a convincing six-person space team.

Then the obvious answer dawned on me: Frank was inspired to become an astronaut by Star Trek. What else could be more natural in this situation but to go for the Star Trek theme? We added one very nice extra touch though. It was the 'Star Trek: The Next Generation' uniforms that the crew of Expedition 21 were wearing in their poster. The commander of the starship USS Enterprise in this series is Captain Jean-Luc Picard from France, played by the British actor Patrick Stewart. Bingo! The first starship captain from Europe became an inspiration for the fun poster of the first European ISS Commander and his crew.

Capt. Picard: "There's still much to do, still so much to learn. Mr La Forge, engage! (last line of Season 1, Star Trek: The Next Generation)

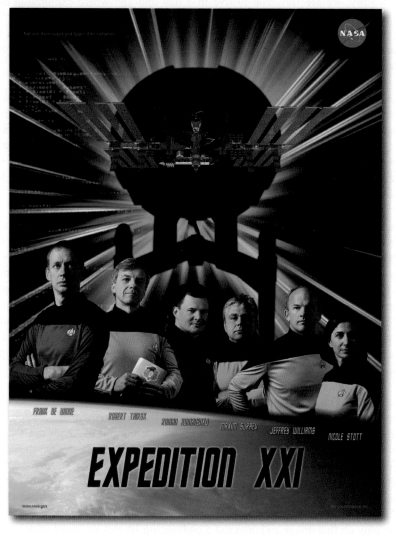

CREW POSTER EXPEDITION 21: CREW IN THE UNIFORM OF "STAR TREK: NEXT GENERATION"

It was impeccable timing because the poster was produced around the time that the latest Star Trek movie was released (late spring 2009). I had a lot of fun shopping for the Star Trek: Next Generation uniforms on Amazon.com. I can't tell you how surprised we were to see the number of NASA trainers who came for the traditional picture taking with the crew in their own Star Trek uniforms! Apparently Frank is not the only one in the space business who found his lifetime professional inspiration in this historic series. I even bought a mask of Data. That wasn't so good. We didn't use it.

As a result of some unexpected events, my own situation at work fortunately turned out in such a way that I could now take the five months leading up to the launch as a sabbatical. From 1 February until 14 May 2009, Frank and I spent every moment together when he was not working. Not only had this given me the time to do more writing, but also to do some fun things for Frank's crew and our family. The special lapel pin with Expedition 21 logo, the tie clip for men and the earrings for women were all the result of the crew welcoming my extended holiday creativity. I'm forever grateful to them for trusting me to do it on their behalf.

I drew the Expedition 21 logo based on the crew's brief that had as its main starting point the significance of the six-person crew. A circle, being the perfect geometrical shape, can be split into six equal parts by its radius and was the basis of the logo shape. The rest came out of combinations of various circles of proportional radii. I was also looking for an original outline that would make the 21 logo stand out visually from all the other expedition logos and could also be used for various items.

FRANK'S MISSION PATCHES FLOATING IN FRONT IN THE STATION WINDOW

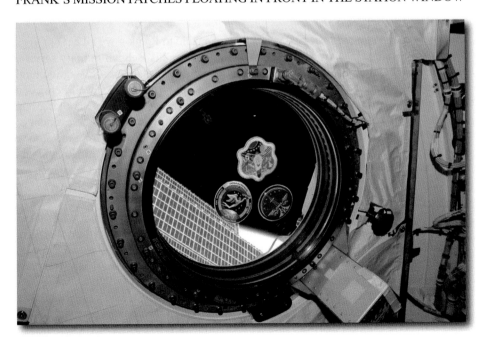

The three parts inside the logo represent the current expeditions to the ISS on Soyuz and Shuttles, the future destination of human space exploration to Moon and Mars. The Moon in a waxing (growing) phase is a symbol of new beginnings. We are living in the Solar System where children are the symbol of the future, for which the results of today's exploration are destined. Sets of six stars represent the current and future crews of six. Fractals of six in the middle are symbolic of the Universe, which consists of the tiniest particles and yet is infinite. Together with the baby-astronaut doll protecting Earth, the seven rays of Sun, the happy kids and bright stars in deep space from the Expedition 21 logo became symbols of Frank's flight.

# Sunny side up

In his past career as a military pilot, Frank spent three months as a detachment commander of a mixed Dutch/Belgian unit in the Kosovo war. There is just one story from those days I want to share with you here. A corporal came to Frank's office with a request for a personal appointment. In military terms, this is a very big and unusual step for a corporal to ask to see the detachment commander, so Frank immediately knew that something serious and important was happening in this person's life. It turned out that the Belgian soldiers couldn't get their breakfast served the way they wanted it: eggs sunny side up, this being the most traditional way of serving eggs in Belgium.

This particular guy felt that he wasn't getting his nutrients because he needed his semi-raw yolk for a successful day. The Dutch cook refused it. Frank went into the kitchen to find out what was going on. It turned out that the Dutch army sanitary rules (unlike Belgian Air Force sanitary rules) dictated that eggs were to be fried in a way that no liquid was left in the dish in order to avoid any chance of salmonella.

But the corporal's request seemed perfectly reasonable to Frank – we all eat our eggs the way we want. This corporal had had them before in his Belgian military service. The detachment was based in Italy, a country with clear standards for agricultural products. As often happens, some rules that were meant for the general well-being of a particular group of people were abruptly interfering in the well-being of another group of people. Frank immediately instructed his medical staff to prepare an assessment of the real risks of eating sunny-side-up eggs in Italy and to propose a solution. It took Frank one and a half days of contacting various top-ranking officials in Holland and Belgium, including the army generals in charge of the operation, to get clearance for cooking eggs in the way that made the corporal feel fit and happy for the day.

The solution found was simple: the eggs could be cooked and served sunny side up as long as they were cooked for no less than two minutes. The cultural differences were respected, the Dutch cook was comfortable when he abided by his rules and a person in need was provided with a gratifying result. Frank took it very seriously when one of the people he was responsible for felt something was a real problem for him. As always, showing the willingness to look for an opportunity where others only see problems, daring to question the rules that didn't take into account personal needs. This brought a solution that was simple, straightforward and satisfying, even though you would have never guessed it was possible.

As I'm writing this story, which I have heard so many times before, I was wondering if it was a coincidence that this particular issue came up for Frank's resolution. Because this is how he is in life: optimistic, positive, straight, and always searching for constructive solutions. Always sunny side up.

## Multi-Culti

Once, when I arrived at home in Holland together with my Belgian family, my only Russian friend in Holland, Ludmila, was already in the house (she has her own keys). We loudly greeted each other as we picked up the conversation in the middle of the sentence where we had to pause it on the phone a few hours back. We were happy because it was a rare opportunity to spend some time together. We talk a lot but we don't get to see each other often. The older you get, the smaller becomes the number of people with whom you can afford let go of your boundaries and be your natural foolish self. Simply because they are only the people from that age when being foolish was a part of being natural. It's almost impossible to develop such relationships when you meet as grown-ups. And the older you get, the more you lose touch with your friends from the past. Thus the number of people who know you who can self-express by running around and telling loudly frivolous jokes and ambiguous witticisms gets exponentially lower on an annual basis. We were shouting and laughing. We've known each other for too long.

As I was enjoying the company of a dear friend, I registered with my peripheral vision and hearing that Frank was trying to convince his Mom that Ludmila and I weren't having a fight and that, on the contrary, we were having a lot of fun. Mom didn't look entirely convinced. At this point, we hit on something outrageously funny and both burst out laughing. This spread a feeling of ease over everyone present.

Many Russians, unless they have a hidden agenda or are afraid of something (which can also appear as being aggressive, because it's common knowledge that offence is the best defence), generally say things as they are. There are plenty of forms of politeness to accompany clear statements to put them light-years apart from being rude or intentionally aggressive. Sometimes directness makes you face some unpleasant situations, but then everything is in the open and you can deal with them and move on. You can compare this in a way to a visit to a dentist. You might not look forward to it, but the problem will only get worse if you keep postponing it. Dealing with it later will cause only more pain. When things aren't even so controversial, why not just say them as they are?

This turns out to be directly opposite for Belgians. The way I feel it, Belgians avoid conflict at all costs. Creating even a mild discomfort in live interactions is seen as very unusual and is almost judged as a symptom of being unwell. Discomfort can be created by calling things by their true names. I see two reasons why this doesn't surprise me at all. One is 'internal': their country is made up of the Flemish, Walloon and German communities, which are in a permanent search for a solution of how to run the country with a population that is smaller than some capital cities. The second is 'external': being the 'de facto' capital of Europe, hosting the gatherings of many nations who get together in search for compromise and consensus, Belgians have taken this up as a part of their national character.

I understand Loredana the best (she's like a sister to me, we just happen to not be biologically related). Whenever a subject is intensely on her mind, she talks to me passionately about it. It's absolutely obvious to me that she needs to get it out of her system. I know that she's not shouting at me but at the subject, we help each other let off steam. It's honest and easy this way. I've realized so many times, by reading between the lines, when people are going in circles trying to make you understand that something needs to be different. This is very burdening and creates a lot of confusion. Or disappointment: sometimes people beat about the bush, trying to avoid doing anything constructive, while spending time making it appear that they are working.

It took Frank and I only a very little time to discover this amusing cultural difference. Nowadays he helps me by checking my writing: if I write anything to a European person or institution, he has a look to see if it's polite in a 'European way' and not only in a 'Russian way'.

During the flight, Frank got even closer than before with Roman, and also got to know Gennadi much better. They didn't have much training together. Naturally they didn't have any time to get to know each other socially.

If you make friends with the Russians, their hospitality remains warm and welcoming no matter where in the world (or out of this world) you are. Work on The Station is to a large extent split between the Russian segment and the Western Segment. Often the crew only get to see each other during meals. Frank soon started visiting Gennadi and Roman for a cup (well, a plastic bag) of tea in the evenings before going to sleep. He would accumulate the things he wanted to take along during the day and carry them with him during this one daily visit that became a nice tradition a few weeks into the flight. Occasionally he would go more than once but that would happen only if there was work to do.

"I forgot my juice there," Frank told me on the phone one evening. "I need to see what else I want to do there before I make the trip."

"Eh? What do you mean 'make the trip'?" I wasn't sure I had heard him correctly.

"I'm in Columbus. I need to go to the other side. It takes time, I should be prepared to do all my tasks while I'm there," he said.

"How long does it take?"

"Depending on the time of the day, the traffic might make a difference."

"Traffic?" I kept questioning the quality of the audio.

MEAL ON-BOARD THE STATION

"Well, if nobody disturbs you, it's between 30 seconds and one minute, depending also if you have something in your hands. But during the day, you might not be able to cross easily because the fitness area is in the way. If somebody is exercising, you'll have to wait for a pause in the exercise so that you cross without disturbing anyone. Then you definitely meet somebody on the way, so you stop for a chat. Then you arrive in the Russian segment, pick up or leave whatever you need and have a small chat there. A round trip can be easily over 10 minutes if everything works out smoothly. So if you go there and back three times per day you spend an extra half an hour. Therefore I prefer to do it no more than once per day because I need those twenty minutes for writing emails and calling you!"

## Multi Culti 2

The biggest cultural difference that Gennadi, Roman and Frank observed on board and shared with me was the interaction between the Mission Control centres and The Station. The communication between the Russian Segment and the Mission Control in Korolev would sound to an inexperienced Western ear as a fierce argument. In reality those were constructive brainstorming meetings between the people who didn't waste time on formal pleasantries and talked about things straight and factually, without spending their time on what would constitute pointless delays in their world. Communication with the Mission Control in Houston followed the more expected

Western ritual that goes along the lines of: "Guys, you're great, everyone here is extremely pleased with your work, keep it up. But if you don't mind, please could you maybe have another look at such and such…"

Frank was fine in maintaining either interaction, as he was pretty much calibrated to receive information in both formats, even though he admitted that the 'Russian way' is more appropriate for operations because you don't want to leave any room for interpretation (this can lead to misinterpretation).

But even being a Belgian, he would have had difficulties decoding Japanese ways. With only 'yes' and plenty of compliments for everything you do, it's almost impossible for a non-Japanese to read between the lines on what is really meant. My experience of the trip to Japan was culturally amusing in this sense. Japanese are tiny. I'm an average person, with shoe size 39 and a European size 40 in clothes. I tried to do some shopping in what looked to me like very special local boutiques. I couldn't find anything that would fit me! If you pick up something to try on which is clearly not your size in a European shop, the shop staff are likely to try to stop you from wasting time on an obviously wrong item and offer you something instead. It came to me by trial and error that I shouldn't waste any time on trying to look at any special Japanese boutiques for anything unique – there was no way I would fit into any of it, and there was no way they would stop me from wasting time trying it on.

In the past, I occasionally had the feeling that Frank was not sure if I was angry or if I was peacefully honest whenever I expressed myself on a subject that was on my mind. By the end of the third month into the flight, he told me: "I know now what you mean when you say that you call things their real names and there is nothing else to it. You're right. It's much easier this way. I now really like it. But you need to live with the Russians to understand and appreciate it."

Thank you Gennadi, thank you Roman, life with you and your pre-sleep cups of tea were the cherry on the cake of Frank's appreciation that 'honest' and 'straight' can easily equate to 'fine' and 'fair'.

## Multi Culti 3

In Russian, there are innumerable ways to change the root word in order to express a huge variety of moods and levels of formality or familiarity in addressing people. This goes for personal first names and for nouns in general. For example, my first name in my passport is Elena. This is a full formal name. Lena is the name you call any Russian Elena until she becomes a grown-up or if you're in an informal situation. I left Moscow when I was 23, and I never became a proper Russian grown-up. For me, Lena remained my name. In Russia we would say Elena is the 'full name' for Lena. Tatiana is the full name for Tanya, Natalia is the full name for Natasha, Alexandra (for a girl) and Alexander (for a boy) are the full names for Sasha. Sofia is the full name for Sonja. In the West, all these (as well as Lena and Elena) are names in their own right.

We have suffixes that express various shades of attitude. The correct word in English for these is 'diminutives'. In Russian, we call them 'diminutive-tender', because you can express affection by adding a particular suffix to a name or a noun. The only example I can think of in English is to say 'doggy' as opposed to just 'dog'. In Russian, if you take my name as an example, you can say Lenochka, Lenulechka, Lenusechka, Lenyska, Lenchik, Lenochek, Lenich or Lenka, and there are more ways. The bottom line is that in Russian, word formation is far more intricate, complicated and, to some extent, more fun especially for foreigners studying it. One of the things we do in Russian is drop the last letter in the name (if it ends with a vowel) when you call someone to draw attention. In English you would say: "Lena, what do you think?" In Russian, if we are on informal terms, you would say: "Len, what do you think?"

Following the same logic, I call Roman 'Rom' when I call him or ask him a question (because Roman is the full name for Roma, and since we are on informal terms, I call him by respective derivatives of the informal first name). This rule works only when you address someone. This form is not used to talk about a person.

Frank speaks good Russian. But even he gets confused by suffixes and derivatives. He registered that, in my informal ways with Roman and Julia, I call them Rom and Yul. Julia is her full name. The informal version is difficult to express in English letters. It's kind of 'Yulya' but in Russian there are fewer letters in the informal version of the name. Now Frank calls Roman 'Rom', Julia 'Yul' and Gennadi 'Gen' as a complete name no matter whether he is addressing them or talking about them. One of those priceless little moments of grace in the realm of intimacy of human friendship.

## The best commander in the world

A crew's numbering and duty allocation is a bit of a complicated story. I'll tell it as short as I can, bear with me. Long-term crews on board The Station have sequential numbers. Gennadi is now commander of Expedition 20. Frank and Roman joined his crew as flight engineers. After Gennadi's departure in October, Frank takes over as The Station commander, the first European to do so, on Expedition 21. Bob, Frank and Roman flew to The Station in Soyuz TMA-15. Roman is the commander of the Soyuz TMA-15 on the way up and down. Frank is a flight engineer on the Soyuz. The other crew members can also be flight engineers, researchers or mission specialists. Space Shuttle missions (called STS plus mission number, STS stands for Space Transportation System) have their own commanders and pilots.

In the months before the departure to Baikonur, we had several parties where it is traditional in Russian to toast with every sip of a drink you take. After the obligatory introductions, and wishing well to everyone present and absent, the line of toasting can start to take its own original course.

"I would like to offer a toast for my commander who is the best commander in the world," began Roman.

"I in turn would like to offer a toast for my commander who is the best commander in the world," toasted Frank in reply.

"My commander is better," said Roman.

"And my commander is better, too."

The first time I witnessed this spontaneous dialogue, I didn't know to laugh or to cry with the overload of sentiment. These two grown men, both military pilots having notched up some serious achievements and survived harrowing experiences, were rejoicing in each other's company like a couple of teenagers who've just passed their high-school exams.

I think we were in Houston when I saw this for the first time. A few weeks later in Moscow when we were together with Julia and this marvellous celebratory chorus started, mostly in English, she looked at me puzzled. I explained to her what kind of ultimate sweet fraternization was going on. Roman and Julia had met when he was a very young pilot. Being the son of a cosmonaut must have put him under pressure being in the same profession. Roman wasn't sure if he wanted to go that way. Julia encouraged him to dare to live his dreams. This brought with it 11 years on the road, with its uncertainties, doubts, ups and downs, hopes and fears. It finally resulted in this six-month separation and who knows what other flights in the future. As we were trying to relax into the fun of the party, with only two of us knowing what was really going on inside our hearts, I had a chance to ask her:

"Don't you wish you never encouraged him to apply to become a cosmonaut?"

"I'm grateful for everything that is given to me. I don't regret anything," Julia replied with the answer I was almost sure I would get from this most graceful and kind person, full of faith.

Throughout the evening Roman and Frank occasionally went up to each other with their glasses filled for a private toast. Secretly they still had water in their glasses, but were impersonating vodka for those around them:

"For you – my best commander!"

"And for you, my best ever commander!"

I can only regret that I never recorded any of these heart-warming exchanges on video.

# Surreal but true

Before the launch, Roman was living at home at Star City with his family. I was travelling with Frank for over four months. Bob and Brenda were separated almost two months before the liftoff when he had to depart for the pre-launch session in Star City. Inevitably I was projecting this situation onto myself and could not even bear to think about this. Not only were Frank and I always together, but also on top of all the great friends I've had for years I got a wonderful new friend in Julia. What Julia and I did for each other simply by being there for each other couldn't have been done by anyone. Only by Brenda. But she couldn't join for the last couple of months before the launch.

A few days after we got home after launch, life had started to take up the shape it would have for the next six months. Brenda and I exchanged an email or two per week, chatting about everything, but more than anything just checking on each other without asking direct questions. I was finding my way back into the corporate world of Amsterdam and trying in vain to keep up my Russian fiction writing. Brenda, as far as I knew was setting their youngest son Aidan up for scout camp.

One month and one day into the mission, Frank and Bob mentioned The Book. They talked me into writing it. It was now up to me to win the support of the other three. I bet you're familiar with the feeling of having to deal with something that you've been contemplating for years and years but never quite dared to. You thought it was just a dream anyway and would never come true. You know you could do it if you had to, and might not actually be bad at it. Yet you are stepping on to new territory, with different rules and undercurrents and minefields. You don't know what might turn out to be a firework and explode in your face. You just take one little step at a time and, in hopeful expectation, observe what happens next. Like a child, with a shimmering candle in uncertain hands, discovering the dark room with a treasure chest in the corner, trying to find the keys to its padlock. A sudden scream! Is that an old broom in the corner or a skeleton?!

I had no idea what to expect. Shared privacy in friendship is one thing. You can expect some help, understanding or even forgiveness in many things. But this doesn't mean people will be happy to explore uncomfortable pursuits. I know Julia is a very deep and introverted person. I know Roman as someone who has a joke for every occasion, but at the same time never says anything about himself. I know Bob as the kindest person imaginable, who can always take a distance and offer a wise perspective on things (but he was already guilty of starting me writing so I took for granted that he agreed to let me write about him). I know my husband, who used to build brick walls around anything concerning his private life. He still does. Now more than ever, he differentiates between human and personal. But he took this as an unexpected opportunity to remind everybody that astronauts are actually humans. One wise man, whose name I don't recall, said that 'The successful man is the average man, focused'. It takes discipline, dedication and focus to succeed in a number of professions. An astronaut is certainly one of them.

I didn't have a feeling at all for what Brenda would think about this Book. Before the flight I had met her only at two or three big events and once when Frank and I were invited to their house for dinner. Over just one night, during which she and her sister Luisa cooked some spectacular Italian food, we all laughed so much that it was clear that no matter how much or how little time we got to spend together, we would enjoy every minute of it.

I needn't have worried. The support I got from each and every one of them was overwhelming. I got to hear, read and write down things that we never got the chance to discuss in all the rush of the docking and return home. Roman was particularly pleased by reading Julia's stories. By sending my draft chapters to our husbands on The Station, I gradually shared with them on 'spouse behalf' the launch and the docking they missed while they were flying the Soyuz.

Beautiful things often happen fast and unexpectedly. It turned out that Brenda is also writing. Her story is about the quarter-century long path that took her to the overheated steppe of Baikonur on that sunny day in May 2009.

We didn't spend much time together before or during the launch, but now not only are we each writing a book but we're also writing pages and pages of emails to each other to share how it's going. Like me, Brenda had been thinking about writing for years. Bob has been encouraging her in every way possible, but somehow it wasn't moving forward. Truly, what goes around comes around. Bob's casual 'you should do it' started an avalanche in my work. In me turning to Brenda for support, this gave her an impulse for returning to earlier drafts. She is an eloquent writer. You have already read a story in this Book contributed by her.

This little story doesn't end there. I was interviewing everyone for their launch and docking experiences. The first thing Brenda said was that her whole experience was surreal. On this defining occasion, she realized that the desire to be near loved ones is universal. I still had her story in a draft form when I got to talk to Bob. No one had seen it.

"Hi Lena, I have about 15 minutes to talk while I'm monitoring the air-to-ground," said Bob, he was on the phone as I was leaving my office to go home. "Everybody else is busy. I just need to pay attention to comms, so I might get distracted occasionally."

I had been trying to talk to Bob for a week. He was saying that we should get his story collected quickly, because his memories of launch and docking were fading away, being replaced by more recent experiences. I quickly ran to my car, pulled out my laptop for a writing surface and found a sheet of paper that was blank on one side and a pencil.

"Baikonur is a different world. It's like nothing you've seen before," started Bob. "But I have to tell you Lena, that whole day of the launch was surreal."

Why was I not surprised?

"Being near the loved ones who came all the way here was such a universal psychological support for me," Bob said, a few sentences later. I didn't want to interrupt him. I'm simply enjoying the thought that when he eventually gets to read all this, it will be a nice little surprise for him. I thought we had only one Romeo and Juliet in the crew. Turns out we have more!

# One other thing you only find out later

I have heard from other cosmonauts that the feeling of weightlessness is very similar to the sensation of falling in your sleep. When I heard this, I immediately knew what they were talking about. Not frequently, but I have had the sensation of falling in my dreams. Frank says he doesn't have dreams. What concerns the feeling of weightlessness itself, he said that only occasionally did it bear a resemblance to falling.

I needed a validation. Roman briefly joined our personal conference just to say 'hi' and I started the conversation with this extremely exciting question:

"How does weightlessness feel to you now?" I asked and followed with the explanation of the falling sensation while asleep that I had heard about.

"I have no idea what are you talking about," said Roman, stunning me with his answer.

"Well, you know, when you sleep everybody occasionally dreams about falling."
"Exactly my point, I have no idea what are you talking about," he said.

I paused as I wasn't sure what to say.

"I always flew in my dreams," continued Roman, to resolve my puzzlement. "Whenever I dreamt that I was in a game or got the chance, I could always fly and be exactly where I needed to be and catch or reach everything I needed to by flying."

"Now everything makes sense," said Frank, joining in the conversation. "Roman is so natural here. It's like he has flown many times before. He can fix himself in the air without making any effort and stay still there. I can't do it, I just keep floating around."

Indeed when I have my conferences with Frank he floats around a bit, even if he doesn't mean to entertain me by hanging upside down. When Roman joined in and I was looking at him while speaking, a couple of times I had to turn to Frank floating and moving around just to be sure that I was watching the live stream, not a still picture (which is what happens when the video link breaks, you're left with the frame of the last transmitted image frozen on your screen). Roman was just fixed, hanging in the air, absolutely still without touching anything around him. Looking at him gave me the feeling of a frozen picture. Roman was born to fly!

Later Frank told me that only as the mission progressed had he realized how his feeling of adapting to weightlessness evolved with time. He caught up with the manoeuvrability he'd learned on his previous first short flight after just a couple of days on board. The muscles 'remember' the earlier experience in the same way that you might pick up a sport in later years that you played as a kid. Ten days into this flight, the adaptation was superbly ahead of the ten-day adaptation for his first flight. Today, exactly on fiftieth day of the mission (Bob calculated that) the feeling of freedom and mastery of flying was liberating. What he had learned was simply amazing. It was impossible to imagine before because there was nothing to compare it to. It was impossible to guess, even with the prior experience of a short mission, what it would be like to 'know without knowing' how to move in weightlessness after almost two months.

On the ground, every step we take is a tiny fall. As you stop walking, you stop falling. In space, Frank learned to fix himself in the air and he didn't fall. I wonder how long it will take for him to relearn to enjoy gravity. I don't even like walking on foot again after just half an hour on roller-blades! The human capability for learning and adaptation yet again proved to be extraordinary. As we were discussing this unexpected twist of personal evolution, we were wondering what other things people might be missing in life because they don't have an opportunity to compare it to a relevant reference. One of the most famous sayings of Søren Kierkegaard, the Danish philosopher, is truly correct: "Life can only be understood backwards but must be lived forwards."

## Staying healthy, staying in touch

Even though The Station lives on GMT, the evenings for me are almost the same as the evenings when Frank was in training in Houston. In Houston, the work finished when his last class was over. On The Station the work doesn't finish. Free time starts when all the planned activities are over. This is after dinner, after the evening planning conference. I haven't gone to bed before one o'clock in Europe since he left. I found peace with the idea that I'll catch up on sleep at the end of November. He is bound to want to sleep a lot when he gets back. I'm just moving through life a bit slower these days.

Sometimes Frank calls just for a few minutes in between tasks. On other occasions, when he has time to chat for longer, I ask for his input in some current unexciting domestic chores. One example was when I had to take our car to be serviced on my own – the first time in my life. I didn't mind, it was actually fun. When you admit to being a helpless idiot, people you've never met before are actually willing to help you. I should remember to do this more often. This doesn't work with people you know because they've already formed an opinion about you. I expect that they don't think I'm an idiot, so asking them for help would require acknowledging some kind of weakness and this is completely different. I just realized that I have no problems with being an idiot, but I'm not able to remain myself and be weak, let alone ask for help, on such a basis or even own up to this out loud.

Anyway, taking our car in for a service on my own will help my ingoing negotiation position when I remind Frank in January next year that his term for uninterrupted bin roll-out has started. I have no illusions: every Tuesday evening he will find all sorts of unbeatable reasons for delaying it with a hidden agenda of not doing it. Starting with the innocent "I'll do it later," or "It can wait until the next collection," going into intermezzo "It's raining now, and who told you it was a man's job anyway?" right up to the sophisticated "I'll just watch the Belgian news now, but I'll do it tomorrow morning, I promise!" What is it with men and the television news? It's all written on the internet anyway! I had to accept the incomprehensible: nothing beats the gourmet dish of watching the news on TV and reading it on the internet at the same time.

Calling from The Station is possible within the coverage zones of Ku-band radio link, which is reasonably regular. It's not possible to call to The Station from Earth unless you work in one of the Mission Control centres and are entitled to communicate with The Station. (Imagine not being able to call your husband for six months? This makes for another creative argument component to keep in your pocket for later strategic household negotiations.) To call from The Station, you have to use one of the computers that are configured to allow the dialling of a phone number. You have to wear a headset. The calls are routed via Mission Control in Houston. There is about a four or five-second delay in the line. Having a passionate debate is completely out of the question. Imagine a heated conversation where you pause for five seconds before you get an answer?! Interrupting doesn't make any sense because it falls into a wrong part of the sentence. You can still make a passionate monologue, but this is rather unrewarding because the line breaks more often than not and you only get the beeps on the line about a minute after it broke. Short complete sentences, to the point, help a lot when you have something to discuss with content. Talking on a regular phone after this is like walking in regular shoes after roller-skating – your body is on familiar territory but your brain is telling you to do something different.

To maintain good health, everyone who flies a long-duration mission has to exercise approximately two hours a day. This is extremely important because the lack of gravity means a complete lack of any physical work. If you think about it, it's obvious. Even if you're not a sporty type, the very fact that you get out of bed and walk around the house means that you are working against gravity. With every step you walk, Newton's third law is working for you: the force in your heels is a reaction to the force of hitting the floor with your foot. No matter how hard you try, it's impossible to avoid some kind of physical exertion on Earth unless you are in bed for extended periods of time. This is how the bed-rest studies are conducted. This is the best known way to simulate on Earth the long-term effects of weightlessness on the human body.

The two hours of exercise include aerobic training and resistance training. The astronauts have a treadmill, a cycle ergometer and different generations of resistance equipment. While running or cycling, they can watch a video. Exercises should be harmoniously combined, each component lasts about 30 minutes. We purposely bought a lot of British comedy series to keep Frank company during his exercises. He particularly likes the British comedy series 'Allo Allo'. British episodes last 30 minutes (unlike American episodes which last 22 minutes). This makes them a perfect company for a space workout.

There is a computer with the phone function near the resistance equipment. Frank calls me on the phone to keep him company when he exercises. Resistance exercises in space are equivalent to working out with weights on the ground. You can't really chit-chat through those though, like you would do with a slow jog. Frank counts the pushes, I try to tell him stories about what is happening in the world in general and in my life in particular. During his endless stays in Houston and Star City, Frank used to listen to the Belgian news on the internet radio. He even bought additional speakers for his laptop for enhancing the quality of the sound. He would have it working in the background in his room as soon as he came from classes. There is no video installation near the resistance equipment on The Station. With the way that the exercises are set up, you need to pay attention to the fine detail of movement and breathing to use your whole body correctly. You can't combine it with stretching your neck in the direction of a screen. But you can have a radio talking at you while you are counting the repetitions of a particular move. And I'm the next best thing in the world after Belgian news internet radio!

## Fifty - Fifty

The launch of the Space Shuttle depends on the weather conditions. Not only in the launch area itself, but also in three different emergency landing locations. The countdown to a Shuttle launch starts a long time in advance but pauses at various critical points of preparation. You can read a lot of interesting details about it on the NASA website.

NASA TV normally gives live coverage of the upcoming launch starting at least a day before. If you follow the news, the forecast of the launch opportunity is regularly given in a form of a percentage or a chance. You would frequently hear numbers like 70 x 30, 90 x 10 or 40 x 60 as the chances for the launch.

Watching NASA TV coverage of the live events can be fun. Occasionally birds or mosquitoes fly close by the camera and then it looks like real science fiction – a huge monster that barely resembles any earthly creature approaching the tiny spaceship. The other day a spider was sitting on the lens for a while. He moved just a little bit in his wandering, he was probably figuring out how to start building a web. The only time I've seen a similar effect was when, during the filming of the first IMAX ISS movie, a stone flew straight towards the camera by accident and cracked the lens. If you watch it in 3D you can see your glasses cracking. Astronauts who've seen it say that the effect in the movie is such that this is the next best feeling to being up there. Have a look, it is fascinating!

Shuttle mission STS-127 was launched on the sixth attempt. Its arrival was very special for us because it was the first Shuttle visit during Frank's mission. Frank called me on the phone shortly before the hatch opening.

"Hold on, I need to catch the laptop, it's flying away," he paused in the middle of our chat. I could see Koichi working on the NASA TV screen.

"I'll go and wave at you, wait," he said and he disappeared for about 10 seconds. He was pointing the camera in the direction of the hatch for the anticipated opening, but there was no one else there.

"Did you see me?" he asked on his return.

"No, I just see Koichi working," I replied. "Oh, hold on. Now I see you entering and waving. Did you go again?"

JAPANESE ASTRONAUT KOICHI WAKATA ON-BOARD THE INTERNATIONAL SPACE STATION

"No! The video delay is about one minute!"

Of course it is. Silly me. Even audio delay is about 5 seconds. Now I could see Frank waving, telling Koichi something and floating away.

"I have to go back there as we have to greet the arriving crew. I'll call you as soon as it's done."

"OK, I'll watch the TV, talk later." We hung up the phones.

As the main crew has assembled to greet the Shuttle crew, they were all facing the hatch and were standing (well, hanging really) with their backs to the camera. It took another minute or so to open. I could see Frank turning away from the hatch a few times, smiling into the camera. I knew he was doing it because we just discussed only a minute ago that I would be watching him. Since the vector of the entire ceremony was pointed in the other direction, he didn't dare to keep turned to me for a long time. He called back as soon as the official greeting was over.

"I saw you pulling faces at me," I said instead of hello as soon as the phone rang.

"I knew you would be looking."

He was extremely happy that our little game worked out just fine.

By the way, while following this Shuttle I found out that there is never a 50 x 50 launch probability forecast. The truth is that 60 x 40 and 50 x 50 in this situation mean pretty much the same, but fifty-fifty sounds too much like a colloquial expression of doubt, or even odds. But they actually do know what they are talking about in the Shuttle control centre.

## To pee or not to pee...

*...that is the question:*
*Whether 'tis nobler in the mind to suffer*
*The slings and arrows of outrageous fortune,*
*Or to take arms against a sea of troubles,*
*And by opposing end them?*

*(Hamlet, Act III, Scene 1)*

This is not really a question. The answer is always 'yes' when you feel you want to, it's much healthier that way. But it's also not a question whether it's noble in one's mind to suffer from the outrageous misfortune of having a broken toilet. You just recognize that if it's not fixed, it can result in a whole sea of troubles. So you have no choice really but to take it in your own arms and, by opposing, end them. Anyway, this would be my creative spin for considering what has happened on The Station in the last couple of days. Fortunately for The Station, the crew thinks about this subject as an engineering problem rather than in Shakespearian terms. Even though it was the 'to pee' answer to the above question that started the trouble.

Remember, we have already said earlier that water is a scarce resource. It comes from, among other things, urine recycling in the American Segment. Naturally, the more you put in, the more potable water comes out of the processor. Now the Shuttle has finally flown (and the crew was doing a great job of installing the new Japanese module), it was decided to benefit The Station's supplies from their physical presence. The Shuttle is equipped to provide life support to the full crew (in the past, Shuttles had performed autonomous missions). When a Shuttle is docked to The Station, all its crews' physiological needs are normally supported by the Shuttle's systems. In other words, the food, bed and hygiene facilities for the Shuttle crew are provided by their own vehicle.

We might not see it that way, but human urine is a valuable resource on The Station. Some of the Shuttle crew were asked to pee in the toilet of the American segment so that more water could be produced.

What wasn't known at that time is that the system was already faulty and didn't safeguard itself from an accidental hit of a button. As a result of this a simple human error occurred. Someone pressed the wrong button. Imagine the stress: you peacefully go to the toilet. Not just 'peacefully', but also a favour for someone (as a matter of fact, can you imagine having a pee as a favour for someone?). But because you press one

wrong button, the whole system blocks up and brings an emergency-level malfunction! It could also happen by accident in flying without even using the toilet. This is stress. And this could have happened to anyone.

The amusing part of it was that, since Frank was among those of The Station crew who played a role in fixing the toilet, the amount of links with articles from friends and comments from my colleagues that I received in these two days was quite incredible. More than one person had come up to me with an expression of condolences on his face. Starting with "I know your husband is having trouble on board because he can't use the toilet," and going as far as sharing serious fear for a jeopardized space mission. There's a lot of truth in the fact that dysfunctional life-support equipment might have a powerful negative impact on a mission. But we were far from that. The first day's operation to remedy any possible emergencies was successfully accomplished by the time the story made it to the big news.

"No," I kept answering to everyone. 'He's having a lot of fun because they are collaborating with the ground while performing some real-time trouble-shooting and looking for new engineering solutions for an operational task." Frank enjoys doing real engineering work (in all fairness, I admit that I do have much easier time at home getting him to look into my misbehaving computer rather than replacing a showerhead or even a toilet roll). The bottom line was that, as expected, people pay more attention to something that is closer to their hearts and homes. In this case, the news about a broken toilet caused far more interest than any news about some successful piece of research.

The toilet break was well timed as it happened shortly before 21 July, the Belgian National Day. It gave a great reason to diversify the Belgian news that day. In response to my explanation of what had happened, three different people came up with the conclusion: "Frank has earned the title of a space plumber." So he did. He laughed a lot when I told him that. He thought it was a fair career step on his way to becoming the next ISS commander.

The best headline of all in this epic story came from Gennadi. He and Frank did a fair part of the work and got a lot of media coverage both in Belgium and Russia. The Russian news item was titled 'Russian Colonel and Belgian General fix American Toilet'.

## 'The Bet'

A couple of months into the flight, Frank started one call with a very specific and untypical question: "Do you know a Russian writer called Chekhov?" I was surprised. I would have thought that for him much more so than for me Chekhov was the Russian crew member of the starship Enterprise from 'Star Trek' rather than a classic writer. Mind, I don't get surprised easily anymore after I heard that there was a general belief that the three Italian logistics modules taken to The Station by the Shuttle were called Leonardo, Rafaello and Donatello in the honour of the cartoon Ninja Turtles.

It turned out that one of the Russian crew shared a story by Anton Chekhov at the dinner table. The story is called 'The Bet'. In this story a very young person agrees to spend 15 years in isolation with all comforts provided in order to win two million after he gets out. Like a lot of writings by Anton Chekhov, the classic Russian writer, this story has several interesting lines of psychological inquiry. For me on the ground, the question of solitary isolation was not even necessarily the first that jumped out as I was discovering his masterful dissection of low human nature. Yet it was too obvious that, for them in six-months isolation (in nice company, mind) from fresh food, fresh air, family and friends, the key question that came to mind, albeit in a jokey fashion, was what makes you accept voluntary confinement, no matter how prestigious and desirable the destination might be? There is a good saying, 'it's like air: when you have it, you don't notice it; when it's gone, you suffocate'. Even though a good breathable atmosphere is provided, and a lot of interesting and much anticipated work is being done, this physical isolation makes you understand the value of own freedom. Freedom of choice, freedom of movement, freedom of doing what you like and sharing with people with whom you choose to make your private life.

If you're still not sure what I'm talking about, imagine that you are offered the job you would love to do for the next six months but the only condition is that you are locked in your own house with five people who are not your friends or family, you're not allowed to go out or open a window, with a limited selection of food, no internet, no shower and only limited phone connection, but with a fully functional surveillance camera. Now, as a mental picture only, slowly lift your house into the air.

# Thinking positive might help

Denial, rationalization, minimization and projection. In modern psychology, the term 'the four musketeers of addiction' was coined for these. When applied specifically to the situation of treating an addict, indeed it's necessary to break through each one in order to move forward. But for everything else, if you ask me, they are a fantastic mechanism that I would call a support mechanism rather than a defense mechanism for dealing with things you can't face.

I gave it a very deep and slow thought and, being a psychologist, irreversibly but intentionally went against the establishment. I consciously decided to live in denial. I focused instead on another key therapeutic premise: don't take away the illusion from a patient if you have nothing better to replace it with. It's nice to help a person learn to walk if they need a crutch. But kicking the crutch away from under his arms won't help if his leg is still healing.

No, I'm not comparing myself to an injured person at all. I don't need any therapy either. At least not until hanging out with your friends and having the occasional shoe-shopping binge is included in the list of officially recognized treatment techniques to be administered by qualified professionals. This is just the way my mind works for explaining a theory by giving a simple example.

Denying and minimizing the disturbance from the idea that my personal life is circling the planet Earth at a speed of twenty-eight thousand kilometres per hour, at an altitude of 400 kilometres and making 16 orbits per day helps enormously.

I made it all by living in denial. Training was a strange fun job my husband was having, and there was nothing else to it. Flying to space was always too far away. That is, until the moment it became too close. Against what is prescribed by my training as a psychologist, I strongly recommend what practicing psychologists would call 'denial for dealing with upcoming stressful periods' – dealing with it in a cognitive plane, as if you are advising a friend how to address rationally something that overwhelms her emotionally, while making it not about your life. Be your own best friend. Give another frame of reference to the upcoming events. Look for fun where sadness is the first natural visitor knocking on your door. Pretend it's not there. Unlike classical psychology, which considers denial a problem, modern spirituality calls it 'being in the now' and praises an ability to achieve it and promises the bliss of a nirvana-like state when you get there. I can honestly tell you, there was no nirvana in the past year of my life and yet everything was just fine.

During week days, I live in denial more successfully. Phone calls don't remind me of mind-boggling relative velocity. Talking on the phone with a few seconds audio delays gives enough therapeutic irritation to distract from the grandeur of the current existential experience. Have you ever noticed how simple things bring you into 'now'? There's another saying that 'if you have problems, wear uncomfortable shoes, then all the other problems will go away'.

# The Third Quarter

As a part of the support to the family of an astronaut, the European Astronaut Centre sets up the video conferencing communication for the family, the support for the launch and the landing trip and tries to take care of various other things that might happen during the mission while the astronaut is in space. In my case, for example, I undertook a major housecleaning at the end of August, including washing all the curtains. The curtain rail was clearly fed up with my bad habit of washing everything too often. Following Murphy's omnipresent law, it decided, for the first time in ten years, to creatively self-express and demonstrate its despisal of my drive for cleanliness while living alone. It collapsed in a way that there was no chance I could repair it on my own.

Now it was opening the view into my living room from the street and made me feel rather unsafe – as if I was living in an aquarium. I didn't dare to call paid help because, unlike in the other specialist work where professional equipment and professional expertise is required, inviting a worker to put in a few bolts to hang up a curtain would be as good as shouting out at the top of my voice "Look at me, I'm living all alone." On top of actually living alone, promoting this information felt too intimidating.

Whereas in Houston being in the family of an astronaut means you are kind of in an extended family of all the astronauts and the astronaut office, life at home in Holland became a life of total isolation from everything that was so intensely my life before the mission. Of course, I knew that I could go and visit (and stay for some days here and there) various people in various locations, but with Frank in space it was not the same. There was no real reason for me to be there.

Then I dared to do something that I wouldn't have thought possible just two or three months ago. I got so much support from my close friends for writing this Book that I decided to ask for another half a year as a sabbatical from my wonderful employers Booz & Company, who have supported me very kindly and professionally throughout this whole year. And I started writing.

I was finally living my dream. I was left alone in peace to do what I wanted to do. And yet I was living on the doorstep of a nightmare: I was completely alone, with all my close people scattered all over the world (varying between a couple of hours driving and half-a-globe flight away or, in one case, extravagantly touring at 400 kilometres above the Earth). When people asked me before the flight, how I was going to cope with this six-month period, I always answered something along the lines of: "It's six months once in my life, it's going to be fine. It would have been a completely different story if it was six months every year."

Now, two and a half months to go until the end of the mission, all of a sudden I wasn't sure how to deal with it anymore. I have been working in various teams for fifteen years. I knew I made a difference and I knew that the results of my work were expected by the others. Teamwork, even though it can create tensions, also offers the dynamics that keep you going. And there I was: my only friend, my only enemy and my only judge of what I was doing. I was literally inseparable from Frank and his life for the last four years and, in particular, every day of the five months leading to the launch. And there I was: in a deafening silence, the clicking of my keyboard and distant trains passing by being the only consistent audio input from the environment (that is if the weather was good enough to leave the windows open).

After only two weeks of working alone at home, all sorts of enlightenments started creeping my way. I realized why solitary imprisonment was the worst kind of punishment; I had to work much harder than I would have ever imagined with the tremendous guilt that sprang out of my heart for having spent too little time with my late grandmother when she was living alone. I had a closer look at the controversial (in terms of consistency of the evidence) and yet well-recorded 'third quarter down' phenomenon in long-term isolation. I discussed it with Frank only in passing, trying to softly ask about his views and feelings, while trying to avoid triggering him thinking about it. It appears that there is no consistent data about this phenomenon in space missions. Thankfully Frank was not experiencing it, but it has been known in other long-term isolations like Antarctic expeditions. Getting into space for most people is a pinnacle of their ultimate dream. The time in space is interrupted by other crews coming and going, cargo vehicles arriving and other special events. There is often something that interrupts the flow of time and prevents it from becoming too monotonous. I'm no expert but it seems to make sense.

I wasn't in the Antarctic but I wasn't in space either. I was alone, with everyone I could to talk to only a phone call away, but yet completely unreachable for interaction in a more humane way. I weeded the whole garden while I was on the phone with my friend Kate Sloujitel in New York. I washed all the windows while I was on the phone with my friend Olga Margevich in Moscow. The three of us have been friends since we were in primary school. Alas, they are two of those few people who made a real difference, both in my life in general and in these months in particular.

I became more aware that I don't understand people who 'live online' in virtual worlds, making friends, having relationships and yet never meeting and not knowing each other's voices, having never shaken hands and never sharing a dinner. The next generation, the so-called 'digital natives' who live this way, turned out to be so much further away than I could have imagined.

There are some things during the flight that came as a surprise, even though I did a lot of thinking in advance trying to visualize a happy and calm six months to come. The fact that Frank can call from The Station made a big difference for staying in touch. It always went without saying that it was my 'main job' in these months to be fully available when he had a chance to call. What I never realized before is that being always available on the phone is a very time-committing place to be. Try it and you'll be surprised. For a couple of months (forget six, it's really tough) always be available to chat with a person who calls you regularly. Never say "Let me call you back," speak with pauses to simulate the few seconds of delay on the line, repeat things several times because the quality of the sound is bad and background noise gets too strong, and never initiate the call yourself – the line is one-way.

On a couple of occasions in the past four years, when there were opportunities for me to take more senior jobs, I had to face my reality. If I did that, we would have been separated much more than we had been already, due to Frank's training. On those few occasions, my moments of decision were extremely short. Our marriage, our togetherness had enough room for flexibility, but we were both prepared not to stretch it. Too often people go for what looks like a feasible compromise at the beginning, until it becomes a habit later, and results in a different life before they know it. I respect those risk-takers, but I realized very early on that I wasn't one of them. Frank and I always looked for ways, no matter how much our life was driven by his schedule, to be together without compromising on the duration and quality of our time. Continuing it the same way during the flight was just natural because we chose to live by sharing our life. From the end of August, with The Book, which was literally happening to me, my total availability for answering the phone, plus a few other writing spin-offs unexpectedly falling into my lap, I started working from home on what may or may not turn out to be a completely new career without having planned it in advance.

My wonderful friend Loredana Bessone told me before, and damn it, as always she was right, that going solo for the likes of us can be a tough call. And yet, I don't know how else I would have made it.

The house started exhibiting all the signs of attention-seeking behaviour. Like a good dog, our house took on the projected idiosyncrasies of the owner. I was ready to shout out at the top of my voice "Come back!" but I didn't because I knew that it wouldn't make a difference. Leaking windows, a damaged floor, a siphoning front door, collapsed curtains, a failed dishwasher, a failed microwave – just to name a few things that happen every few years in a regular household (and not all at the same time). But they all happened to me within a period of two weeks. The world was becoming weird.

I was trying to follow the advice of a few close friends that, in order to rebalance my new reality, I should create some physical routine. I thought 'labour therapy' was a better alternative to a fitness programme. Doing the garden is hard physical work. It has more use than just running or jumping. In the front of our house there is a strip of land that I bet has been called a jungle more than once by passers-by. I decided that clearing the energy of the house and the garden, and combining it with physical activity, would be the way to go. I started by taking a fresh and daring decision – to learn to operate an electrical jig-saw.

Everything that could have been done by hand I had already done a couple of weeks ago (on the phone with Kate). Now it was time to do a real trim of the old overgrown 'evergreens'. I always wondered why they were called evergreens when they dry up and go disgustingly brown. The only difference in my garden being that they don't lose the brown bits naturally but stay untidy, look unwell and depressing. With my general belief in good energy and residual recollection of Feng Shui, in which I took a short but intense interest over a decade ago, I decided to cut the evergreens into a fresh shape. That bush was really getting on my nerves. For once I'm not talking about a badly waxed bikini line or some silly Texan fellow.

After a few unsuccessful attempts to get the jig-saw going, I finally found the combination of buttons that allowed it to be turned on. Fortunately for the world, it was well safeguarded for experimenters without decent muscle strength and experience. After about fifteen minutes of practicing on smaller growths in the front yard, I was feeling like Arnie Schwarzenegger mining for plutonium on a remote planet in some science-fiction movie (thankfully I didn't quite reach chain-saw massacre standard). My senses caught up with reality: my hands were shaking and my shoulders were hurting. Then the obvious happened – I cut the power cord.

"Let's take that as a sign from the Universe that it's time to pause and to revisit your skills of an electrical engineer with a full master's degree and zero practical experience," I said to myself, while trying to fill up my time with a positive attitude to unexpected challenges and opportunities. With a deep resolution to fix the power cord, I went into my living room. No light, no internet, no phone, apart from the mobile. Not only had I ruined a power cord, I had created a short circuit that blew a ceramic fuse. With my tail between my legs, I had to call someone at work to help me figure out what to do, since flicking all my fuse switches didn't make any difference. New challenge: go and find a 16 amp ceramic fuse. Oh my God! I thought they grew in boxes inside power meter cupboards!

Shortly before all this happened, another person became the new acting Head of the ESA Astronaut Division. He took my little curtain dilemma very close to his heart and the next day his colleague came by and fixed it in half an hour – as many men would be able to do.

We had only met briefly before on a couple of occasions. When he was given this management position, he said he wanted to make it his personal responsibility to make sure that Frank and I were cared for as people, with our feelings and considerations taken into account wherever possible, both during and after the flight. For the first time in many months, I had an illusion of being safe and not judged. I was very lonely and my usual alerts were off. It seemed like a good idea at the time to talk openly about how much I missed my husband, how difficult it turned out to live in isolation, with all the families and my close friends in different countries. The truth was that there was almost nobody nearby where I could turn up on a door step unannounced, and be as happy or as miserable as I felt, without having to put up the performance expected from the humble wife of a hero.

At this point, I had been almost breaking into tears when people asked me casually, "How are you?" and even more so "How is Frank?" The only answer I had was, "Too far, for too long." But this is not what was meant by the question. If, in the past, I cried easily over sentimental movie endings, by the fourth month of the mission I had reached the point where I cried when people showed unexpected and unsolicited acts of kindness: as little as a spontaneous phone call and expression of an unexpected interest in my well-being. I was used to being impeccably logistically handled in a manner prescribed by the system that was designed to tolerate no emotion. I have learned by experience that it was my having feelings and expressing them that was discomforting to the system.

NICOLE STOTT AT HALLOWEEN. MULTI-LAYER INSULATOR OF THE JAPANESE CARGO VEHICLE HTV KEEPS THE PARTY COOL.

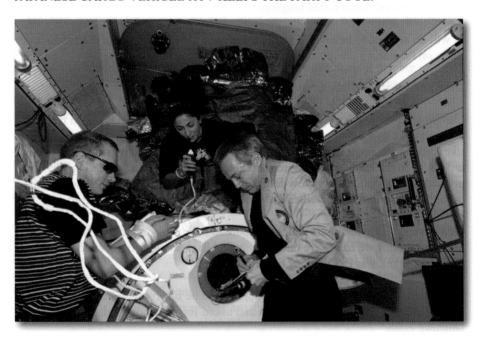

TEAMWORK DURING HALLOWEEN WHILE FRANK IS "TIED UP"

In order to answer a call from Mission Control, you have to press some button on board. Communication is one way. When The Station talks, Mission Control has to listen. Only when The Station lets go of the 'talk' button, can you talk from Mission Control, and vice versa.

And there they were: Houston to Frank. Only now he'd just been tied up around his wrists. Team spirit truly works miracles. One of the crew immediately found an excellent solution:

"Hello, Houston. Frank is, ummm, tied up at the moment. Can I take a message?"

"Please ask him to come on the line," said Houston sternly.

"OK, he will. He just got tied up," it was too good not to use again.

"We will hold for Frank."

"Sure. Frank, come on!"

The six people managing to control their laughter were real heroes. Have you ever tried to remain serious when everyone is laughing, or feel genuinely cheerful when every single person around you is sad?
"Frank here," he appeared, holding the microphone with his tied hands, keeping a straight face. Yet again, the truth and nothing but the truth had been told to the ground. Notes were taken down by a crewmate, the button on the microphone was pressed by another crewmate.

"I am here. It's OK now to take notes."

In the background, the last remaining orange from the previous Progress supply vessel had been slowly making its way towards a ventilation inlet, proudly impersonating a pumpkin with a scarecrow-face cut in it, ominously glaring white flashy teeth. One of the guys was taking pictures as this was happening, but was laughing so much that he couldn't operate the camera.

## The world lasts because it laughs

I remember one of the topics in my English class at school. I even volunteered to draw a poster, which then hung on our wall for a couple of months. "The World Lasts Because it Laughs," I wrote in huge red letters. It was not necessarily logical to substantiate this overall title by offering a class on Liza Doolittle and paradoxes by Oscar Wilde. Yet I tend to agree with the general sentiment. If you can laugh at a monster, all of a sudden it's not so scary anymore. If you can laugh at yourself, and maintain a realistic view of yourself, people will feel at ease and comfortable around you.

It's hard to tell if that series of English language classes at the age of 14 left an imprint on my general perception. For the first time in my grown-up years, I thought about it as I was finishing my psychology thesis proposal: 'The correlation between the personality type and the sense of humour'. Apart from the resulting PhD in psychology, my inability to laugh at jokes for a couple of years was due to the fact that my mind was automatically dissecting them for structural analysis and some interesting, but not particularly useful, knowledge about Freud's ideas on humour and the subconscious. I arrived at the end of my studies with an even deeper conviction that the best way to study, talk, interact, engage, support or, in other words, live, is with a joke. "Blessed is he who has learned to laugh at himself, for he shall never cease to be entertained," remains my favourite quote from an anonymous wise person. Fair enough, he or she might not have had an easy life, in order for them to discover this inner solution.

What does it mean in my life in practical terms? When normal people read their morning news, I read the current new edition of online Russian jokes. Those who read the news write jokes about them, so I'm not actually missing anything really important this way. Besides, there is no internet on board. In a moment of absolute clarity, just a couple of days into the mission, I realized that this is the best daily support I could provide. For six months, my morning routine during the mission consisted of, after brushing my teeth, copying fresh editions of Russian jokes for emailing to The Station.

## Save and Protect

A week before the landing, I joined my brother and sister-in-law Alex and Tanya in visiting their friends Oleg and Natasha, who have a lovely house along Kashirskoe highway, quite close to Moscow. In the misty darkness of a Russian late November evening, all our attention was focused on the bumpy road and we didn't get to witness with our own eyes the ultimate safety masterpiece of the region. This was a big black cross, right on the worst bend of a narrow country road just off the highway. Many people had bad accidents there when they missed the turn and couldn't control their cars. Reflecting markings are not part of the Russian traffic regulations. When the local residents discussed privately installing posts with shiny stripes along that part of the road, they realized that it would be a complete waste because the authorities would soon remove them anyway.

Finally, instead of installing a proper sign-post to warn people about the imminent deathtrap on the road, someone installed the massive black stone cross with the prime prayer to the Saviour: 'Save and Protect'.

Accidents keep happening at the same rate. But now the locals have this real Russian story to tell to their visitors. It always comes with a warning. "Don't try finding it in the dark, just be warned that there are nasty turns and watch the road." Why does everything sound so painfully familiar again and again? Why do Russians find it funny? Why does everyone else carefully nod while not quite understanding what I'm talking about?

Why not wake up to the reality of today and make things better? Why pretend to solve a problem while wasting resources on hypocritically reinforcing things that don't work? Why persist in investing in symbolism instead of addressing the core issue? Why fiercely fight for the form and deviate from the essence? Where does this self-mutilating, obsessive Russian love come from for creating and substantiating problems when easy solutions can be found?

With a 'minor' giggle that I have acquired in the past few months, as a reaction to being occasionally managed instead of supported, I listened to this little story from our family friends. Why wasn't I surprised that this story brought me back for a few brief moments to that brick wall of ice-cold rejection from the management in charge of the emotional well-being of our crew and their families?

# 1 Landing

## Me

A painfully real countdown started for me at the end of August, when there was half of the mission to go (minus one day) and the number of the remaining days went into two digits – below one hundred. At the end of September, the number of remaining weeks went to one digit – nine weeks to go. At 42 days before the landing, the number of hours remaining until the landing went down into only three digits (below 999), and at about four days before landing, the number of hours went below two digits: 99.

No, I didn't become obsessive-compulsive with numbers. It's only a figure of speech that I was counting hours. These homely mathematics entered my life from the stories I was hearing often enough from on board. They were also counting down. Yes, they love their work. But yes, they're the new generation achievers who love and miss their families and have no problems proudly embracing the fact that they're looking forward to coming back and returning to normal lives. It's great to have flown. It's even greater to come back home. It's thrilling to look forward to an opportunity of yet another flight.

Throughout the final pre-landing period, people around us yet again didn't cease to amaze us. They were repeatedly surprised that Frank wanted me to meet him at the airfield in Kazakhstan and have me on the plane with him on the way back to Star City. "Why does she need to be there? It's only a difference of a couple of hours," they kept saying.

Is it really possible that there is a vast shortage of people in the human spaceflight business who share their very special experiences in life with their significant others?! No matter how many times in the past year that Frank expressed this concept – that the flight was a shared experience for us where physical separation was a mere inconvenience – only very few people seem to have been able to grasp it. Amazing.

I got a genuine offer to go with our whole family for a dinner event a few minutes after the arrival of the crew at Star City from the landing site: "He will be tired, you might like to spend the evening at the official reception, but you will not have to cook though." I had to hold myself from bursting out laughing when I heard this. I was proud to have found the right words to explain that, for Frank, first of all it would be important to have his mother, sister, wife and kids around him who, in turn, don't mind to do some extra cooking, as long as Frank is happy with how things are going. Frank, as expected, was very amused when I told him this story.

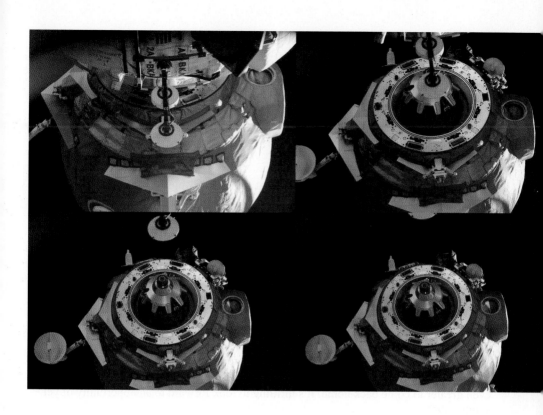

LANDING

"Management will come to greet the crew. If the family is with Frank until this point, please leave all but the wife, so that the management does not have to see them," was the next cultural shock we were addressed with, and to my great surprise more than once. This is still after Frank has expressed to various people in charge that he wanted his family with him while the others were welcome to visit. Isn't the management of the international human spaceflight programme carried out by humans?! What can possibly be wrong with greeting an astronaut who happens to be happy with his family around? What do they possibly hope to achieve by antagonizing the family? Oh yes, I forgot, the psychological support and emotional well-being, which is known to be important. Guess what, management came and greeted Frank. They also greeted his family and toasted to his mother.

Less than a week after the landing, all this didn't matter anymore and only stayed in my memory as a set of anecdotes about some peculiar cultural quirks. They felt so nonsensical now that they don't even seem to be about my life. And yet, when I think about it, why can't I help but recall a paragraph from an introductory psychology book, that confident, established and fulfilled people don't put others down, especially those in weaker dependant positions? Why can't my mind help drawing parallels? Alas, I hope the system catches up sooner rather than later with the fact that we are living in the 21st century. I hope the others coming after us have an easier life and get real human support, especially in the emotionally critical moments like the launch and the landing.

# Him

On reentry into the atmosphere, the weight of your own body that you haven't felt for six months starts in one fraction of a second to weigh you down with a heavy load and feels as if someone big is sitting on your chest.

And then, shortly before the touchdown, the final parachute opens. It sends the capsule into an uncontrolled jerk that for many people is extremely physiologically tiring. Frank knew what it felt like. This is a critical moment, important for a safe landing but at the same time giving Frank the most uncomfortable sensation. He was not looking forward to it. But not for Roman!

"Wah-hey! This is much better than any Six Flags ride I've been on," was the spontaneous reaction of this natural-born space-flier, who had dreams of flying in his sleep all his life.

"Blood is a powerful matter" whispered my subconsciousness the words of Voland from the Russian classic Master and Margarita as I was listening to the story of this second generation cosmonaut. Bob just remained super-cool and reassured his crewmates that he was fine and happy they were safely approaching Earth.

The capsule touched down perfectly – softly and upright. This doesn't happen too often. This is a huge advantage for the comfort of the crew and their evacuation. The weather was terrible though, none of the rescue helicopters could get off the ground. The crew was picked up by some multi-terrain vehicles. These ancient but reliable monsters were slow. When Frank finally got to call me, his voice was shaking and dull from the bumpy cross-country ride over the confusing, cold, yellow terrain of the steppe. If I didn't know better, I would have guessed he had a sore throat and that he'd had something nice to drink. Finally, the cortège reached an area with mobile phone coverage.

I'm rather confused with the speed dial function on my mobile phone. On my Moscow phone, Frank's number is programmed as 'Aaaaa', to be the first one in the contacts list. 'Aaaaa' flashed on my phone as it impatiently screamed at me with its 'ring-ring' (the ringtone called 'nostalgie' in the tone list). YES! For the first time in over six months, I can not only pick up the phone, but also call my husband! They were close to Arkalyk and would soon be in a hotel.

A large number of things are different on the return from a long spaceflight. First of all, you lose this magnificent ability to leave things hanging in the air next to you when you need to do something else for a second. As everybody else in this position has done before, Frank dropped a fork as he went to reach out for a piece of bread. Your shoes weigh you down like tubs of cement in Italian mafia movies, your shoulders ache because you have to carry your head again after it has been totally weightless for six months. Your skin aches from sitting or lying because in space, even strapped into a sleeping bag, you restrict yourself from floating but you still don't rest on any surfaces heavily. The same for the clothes – they are made to fit you but, unlike on the ground, you hardly rub against them inside. This means that your body hair grows longer. I've never thought of it before but body hair does rub off and change in thickness and length depending on the tightness of your clothes and your lifestyle. The crew doesn't wear any boots apart from those for exercise. As a result of six months in weightlessness, your feet become extremely soft, like the feet of a baby who's never walked, and your skin becomes very sensitive. The dead skin cells don't naturally exfoliate as they do on the ground. After your first warm bath on Earth, your skin starts falling off very intensely.

The next obstacle in the return to Star City was that it was too late in the evening to leave that same night from Arkalyk to the airport in Kustanai for the flight to Moscow. The first night on the ground was passed in a mediocre hotel, which was decent by the local standards but yet something altogether different compared to what you would expect when returning from a six-month spaceflight.

When we first heard the news that our husbands would get back to Star City only a day later, Julia and I turned to each other and, almost at the same time, said something that we were both avoiding facing before. The landing was delayed from the originally scheduled date of 23/11/2009 to 1/12/2009: $1 + 1 + 2 + 2 + 9 = 15 = 1 + 5 = 6$. Six and four are not the best numbers in numerology. That is if you believe it. Avoiding to talk about this in the past months was as good as disbelieving it and making it not real. There is a good chance it's not real. Delay in the return was not important. The only important thing was that they were safe, healthy and the landing was soft. In the reference system of flying into space that is.

The next morning they were still not sure if it would be possible to travel due to the weather, but eventually it worked out. Only a three-and-a-half-hour flight before we are together.

Finally the crew walked off the plane. Frank was looking for the family in the crowd. As he found me with his eyes, he sparked up. Before reaching us, he had to go through a row of some big Russian bosses, shaking hands and holding him up with their conversation. And finally he put his arms around me. We pulled into one big hug with his kids standing nearby. We were just laughing as people do when they are truly happy.

All of a sudden, all those pains, frustrations and fears didn't matter anymore. It didn't matter who said what, or which of my expectations turned into heart-tearing disappointments that will still take time to heal. It didn't matter anymore who thinks what about me, or who was uncomfortable with Frank choosing the company of his family over bureaucratic glamour games, which are sustained under the auspices of Soviet traditions and seem mostly to be used for validating egos in the system. It didn't matter that, a few moments before the plane landed, I was told to go together with the kids and sit in the bus to wait for Frank, instead of meeting him at the bottom of the staircase of the plane. I didn't do it anyway. Peace and warmth to all. Work out of love, and the world will smile back at you.

LEFT TO RIGHT: BOB, FRANK AND ROMAN THE DAY AFTER THE LANDING

# Us

On the evening of the crew's return to Star City, as long as I wasn't touching him, Frank looked to me as if he was someone else awkwardly wearing my husband's body. As often happens with my complicated imagery, he found this explanation funny but wasn't quite sure what was I talking about. It took all of us who were there, Tom, Nele, Koen, Jeanne, Carine, Jack and myself, to hold him on his return just a few seconds to get back into our usual ways of having him around, talking about everything in the world, sharing good laughs and hearty hugs, and yet I couldn't let go of the sense of his fragility caused by the way he moved.

I was so used to being alone in the last six and a half months that I couldn't sleep that night. It was so precious to be together again that I didn't want to fall asleep. I was afraid to miss any moment of this magic dream of having him back. I was so overwhelmed by events that I couldn't sleep anyway. I knew though that I was overspending my body, as had happened already so often in the last eight months. I had to surrender to sense and took a very light pill, which just helps falling asleep – the only way I was managing to have any decent sleep in the last few weeks. I fell through the surface of reality, still not quite believing that the new phase of our lives had already begun.

The next evening, now over 48 hours after the landing, Frank looked to me almost normal. Awkwardness in movement was slowly dissipating, even though I could see that he was still concentrating on holding his head straight. He was feeling muscle pains all over his body because now moving around in gravity was one big work-out.

When we were finally alone that evening and quietly sat together, Frank laughed as he could see in my eyes some residual disbelief. Was it really him here in the room? Or was my disturbed imagination creating images of my husband in some erratic and desperate dream to pretend that I'm living a normal life? We were talking all night. We could both no longer comprehend that the whole six months were behind us. It was too big and it was too insignificant. It was too magic and it was too painful. It was the biggest adventure of a lifetime, and it was the toughest and the most exhausting challenge of separation we would have never wished upon ourselves. Our year 2009 now existed in two randomly oscillating dimensions: it has been forever since we were together. But it seemed only just the day before yesterday that we stole a couple of precious hours of intimacy as Frank's colleagues took the risk of 'smuggling' me into their room in Baikonur, in that daring attempt to protect us from the ridiculous pomposity of the inhuman treatment preceding a human launch to space. We were finally together again after the unbearably long divide that felt as if it would never end. We haven't been apart for more than a fraction of a second since we met a few years ago.

The central part of this most magic and most trying adventure – the spaceflight – is over. Just a week after the landing, I could almost sleep normally and Frank was almost not hurting. In happy disbelief, we were standing almost firmly on the ground as winter finally hit Star City on 7 December. A snowfall had started as the formal part of the Vstrecha was coming to an end in the early afternoon.

We're not sure exactly what is going to be next. But one thing we do know is that we're looking forward to living a normal, easy, fun life and sharing it with family and close friends. After all, it's common knowledge that life is what you make it.

# Vstrecha

There is a traditional, celebratory meeting held in Star City about a week after the return of every crew from their flight. Everyone gathers near the Gagarin memorial, where the crew lays flowers and receives congratulations from the officials. Then the whole procession walks to Dom Kosmonavtov to join the official meeting, where the crew and a lot of space officials sit on the stage and share a couple of hours of speeches in celebration of the return of this particular crew and of human spaceflight endeavours in general. The formal part is followed by a reception for the officials, personal guests and families. This is what Frank had to say on this very special occasion:

Dear friends,

I would like to thank you all for sharing this special moment with us today. Vstrecha is a unique occasion which signifies the end of the mission and a new beginning – it opens a new chapter in life after the flight.

A lot of good words have already been said from this podium today about the achievements during this flight. Indeed we had the privilege to be part of a number of 'firsts': as we arrived, we were for the first time with all the international partners present on board the ISS, Russia, USA, Canada, Japan and Europe, augmenting the permanent crew to six. We have received the first Japanese HTV. For the first time during our mission, there were three Soyuz vehicles docked to The Station at the same time. For the first time, there was a European commander on board, for the first time there was a Canadian on a long-term expedition. For the first time, the commandership on board the ISS was handed over in the presence of a Shuttle crew.

I have always been proud of being a European; and Europe has a big heritage of exploration. Therefore it was a particular honour for me to represent Europe as the first European commander of the International Space Station, today's only exploration outpost in space. It is also an honour and an achievement of the European Astronaut Corps and of the ESA Human Spaceflight Programme. It clearly shows what we can accomplish together as Europeans. It also symbolizes the success of the International Space Station in which all partners play an important role. The ISS is a true example of what humans can achieve when they decide to work together for a common goal leaving aside their differences.

I strongly believe that a society that stops exploring is a society that stops progressing. Therefore, I hope that Europe also in the future will continue to explore and take up more and more responsibilities in space exploration, including manned transportation capabilities. European ships and sailors sailed all the oceans of our planet. I hope to see European manned spaceships travel to the ISS and beyond.

Our joint intellectual efforts multiplied by the human spirit of exploration will take us to the new heights.

A lot of praises and congratulations were said here today to us, the crew. Indeed we have worked hard and we are proud of what we have achieved. But we know very well that this was only possible because of the role the other people played in this mission. I would like to thank everybody in all the partner agencies who designed The Station, our mission, implemented mission control and crew and family support. Our flight is a tip of a huge iceberg, which is strong and powerful because of the grandeur of its invisible part. Thank you, all. And especial thank you to the instructors here in Star City who became an extended part of the crew for us.

Like every big achievement in life it comes at a price. And sometimes this price has to be paid, not only by yourself, but also by the people close to you, those you love the most. Long-term spaceflight is a marathon not only for the crew, but also for the families: wives, kids, and parents. It requires strength and resilience from us, the crew, but also from our wives who build their lives around our careers and dedicate all their time, attention and effort to supporting us. For our wives, much more so than for us, these months of separation which started in the first half of May and ended at the beginning of December were a time not only of great pride for us but also a time of hard work; in many ways, it was much harder for them than for us as they had to stay in the shade and to adjust their lives to all our professional needs, changes of our plans, and impositions of the requirements by the others.

I believe that human relationships, human kindness and human attention are extremely important in human space exploration. We take our human values and go out into outer space in our human quest to share them, as we build our human presence outside the boundaries of our planet Earth. I would encourage everyone engaged in human space exploration to always remember that 'human' is the first word in this expression.

Brenda, Julia, Lena, it is your love, patience, understanding and support throughout the years that have made this mission successful. Only thanks to you, have we achieved the success that is celebrated here today. A lot of flowers and gifts are given to us here on stage today. As far as I am concerned, those are for you. We love you.

LEFT TO RIGHT: BRENDA, LENA, JULIA AFTER VSTRECHA

# And life goes on

As I was finishing these last words of the landing story after the Vstrecha last night, Frank was reading his email: "Tomorrow there is a dinner in Cottage 3."

"We definitely have to go," was my instant reaction.

I haven't left Star City yet but I'm already nostalgic about this tough but incredible life we're going to leave behind now the flight is over. We definitely have to go to this dinner because this would be krainii dinner for us and for Bob, but especially because tomorrow our friends, astronauts Soichi Noguchi and TJ Creamer, depart for Baikonur for their launch to the ISS on 21 December. We'll go tomorrow morning to wish them well at the traditional formal breakfast. I will silently pray for their warmth and well-being. It is winter out there.

EXPEDITION 20 CREW REUNION AFTER THE FLIGHT. LEFT TO RIGHT: GENNADI, MIKE, KOICHI, ROMAN, BOB, FRANK

# A few closing thoughts

## Space, microscopes and future cosmonauts

Pretty much all Russian children born within twenty years after Gagarin's flight at some point in their childhood wanted to become cosmonauts. I got over this by the time I was in the middle of the first grade at school. Then being a teacher seemed much cooler. Then a doctor. Then I grew up, but that is entirely another matter. Of course, the profession of cosmonaut still remains very special and prestigious to the present day. But these days, you would hardly find any real interest from the general public in any details of space conquest. It seems after the Soviet time, the majority of the country's population still holds on to the deep belief that in space, circuses and ballet, at least, we are well ahead of the rest of the world. This contributes to the overall sense of inner well-being and doesn't need looking into any further.

By the time we were taught to write, the first task came: write a composition about 'What do I want to do when I grow up'. Naturally, the text of the composition was not expected to be much longer than the title itself. What and why in one sentence was the scope of the task. I wrote: "When I grow up I want to be a cosmonaut and discover new planets which are not seen via a microscope."

Schools in Russia were in those days organized so that, from the very beginning of primary school up to final completion of secondary school, you would go to the same school. Graduation from school is signified by two major events. One is called 'The Last School Bell'. This happens at the end of May after the last day of lessons. Since a bell rings to mark the start and the end of every lesson, this symbolic 'Last School Bell' marks the end of going into classes. Traditionally, the students who are just about to graduate get together with the kids who are finishing their first year in schools. The first-graders prepare some performance, typically reciting some poems for an inspired future and give flowers to the graduates. When the poetic greetings of the kids and speeches of the teachers are finished, one of the kids rings a big bell, and each graduate takes a kid by the hand and the whole procession walks through the school where the rest of the students are waiting to bid farewell. The graduates then take exams in June and have a final graduation ball at the end of June.

When I was finishing the first grade, I was among the candidates to be part of the recital of inspirational poems at the 'Last School Bell' event in my school. The poem read:

> Представить трудно, что случится завтра,
> Быть может в космос полетишь, вот ты,

> Тебе как будущему космонавту
> Я первая хочу вручить цветы.
>
> It is hard to imagine what will happen tomorrow
> May be you would be the one to fly to space
> I would like to be the first one to give flowers
> To you, the future cosmonaut.

Alas, my reciting abilities were not grand. I never got to read that poem at the ceremony. But I knew all along there was a reason why I spent the whole day before the selection rehearsing it. So that I can recite it to you now.

# A shooting star

Russians are full of superstitions. Starting with the obvious ones, about black cats crossing the road (and, just in case, all cats crossing the roads too), up to the vast unexplored realms of individual peculiarity, which are deeply rooted in ancient paganism. Women with yokes, gypsies in shawls, owls at dawn, crossing the door of your office with the right foot, you name it, we will tell you what it means and what has to be done to counteract the negative consequences of experiencing it. I personally used to practise two positive superstitions. One is that you can make a wish on a shooting star and it will come true. The other one is that you can make a wish on New Year's Eve as the Kremlin clock on the Spasskaya Tower goes ding-dong, and that will also come true. I still do the New Year wish, but it becomes gradually more and more difficult because I need to remember the time zone I'm in for a given New Year's Eve. So I do it, just in case, on the hour for several hours – just to be sure. For the last ten years, I've defined my wish in broad terms of world harmony. Even though I don't see many positive results, I cherish the belief that thinking about it regularly can only help the world.

If you've not looked into this subject before, you might find it astonishing that you can see The Station with the naked eye. Websites of the partner space agencies give links to the visibility times in your particular region. The Station has to be flying somewhere over your head in the dark part of the day for you to be able to see it. And of course there should be no clouds. The space agencies don't forecast this bit on their ISS spotting links. Of course, the space agencies build, launch and operate meteorological satellites, but this is an entirely different subject. The Station flies in a low Earth orbit, but Earth also turns so the position of The Station in relation to various points on Earth keeps changing. You can see The Station (and also satellites) the same as you can see the stars.

I know this now. I didn't know any of this 20 years ago. I didn't even know that the Russian Mir space station existed. I just about knew that the new rockets were called Soyuz and not Vostok or Voskhod (I was confused about those as much as I was confused about which dog was Laika, which one was Belka and which one was Strelka). Along with the rest of the younger Russian population (with the exception of those working in space or reporting about it) my general sense of superior well-being was deeply satisfied by the broad knowledge that we led the world in space, circuses and ballet. Since I happened not to be interested in any of those three, it never crossed my mind to find out any further details.

I was a young romantic who studied electrical engineering and loved, more than anything in the world, my dog Art (named after Oscar Wilde's Art for Art's Sake). Art was a 'Lassie' collie, black and white with just a few streaks of red. He was not particularly bright (this is painful to admit when it's common knowledge that dogs take after their owners) but I found gratification in the fact that he was stunningly gorgeous. Dear feminists, please don't throw rotten eggs at me, but he was living a wrong incarnation. He could have been a perfect woman: kind, tender, faithful, thin, tall, beautiful, sensitive and responsive, with a tiny imperfection that made him even more adorable, and with an uncomplicated thought process.

Dog lovers would understand me, others shouldn't even try. We had a good time during our evening walks. I would talk to him about the meaning of life. He would ignore me, as men do. In winter, when small pieces of ice would freeze between his toes, he would come to me and put his paw into my hands, "Warm me, I am in pain." He used to stand on his back legs, put his front legs on my shoulders and bite the edges of my ear lobes when he was particularly happy. He didn't like me brushing him. He thought my hand had suddenly grown a disgusting scratchy extension. He tried to bite the brush off my hands. In the late 1980s, in my engineering schools, you had to make hand-drawings. Whenever I put a board on the floor and pinned A1 size paper to it, he would come and lay in the middle of that paper, so that he could be part of what I was doing.

For some reason in August in particular, there are many shooting stars over Moscow. I'm not sure if this is a global phenomenon – I've never met a non-Muscovite who would possess this sacred knowledge. Maybe it depends on the climate? In Russian, they are called literally 'falling stars'. Maybe they are riper in Moscow at the end of summer? Anyway, it was on one of those long Arty walks, that I was staring into the sky and he was trying to make friends with a cat. My Dog was not informed at birth that he wasn't supposed to like cats and was actually able to swim. I stopped trying to take him with me to the beach after he almost had a heart attack watching me going into the water. In my childish cruelty, I thought that he would eventually join, so I stayed in for quite some time. Poor thing, he ran up and down the stretch along the beach barking at the top of his voice. He tried to enter the pond a couple of times, but could not make himself go any further as soon as the water touched his belly. When I eventually got out, he ran and jumped at me, dropping me on the sand and spent ten minutes finding his place near me, not quite able to settle down. We eventually compromised with me sitting and him lying with his long nose stretched over my lap.

Getting back to that walk, when I was staring into a perfect sky, full of stars. The end of August always makes me a bit sad. I love long days. August is a reminder that they are possible but will be gone very soon. Then I finally saw the star I thought I was looking for. But it wasn't shooting or falling, it was moving steadily and constant. It was very different to the sharp kind of downward move that shooting stars make. I made a wish just in case. I have no idea now what I wished for. The trajectory of that shot left me puzzled and it made me wonder. Only over a decade later, I realized that maybe I had made a wish that night on the coincidentally passing Mir space station.

Today there are websites that contain tracking software, you enter your geographic location and get the ten-day forecast of The Station visibility for your region. Roman and Julia did this in a more personal way. Once when Roman was checking his own navigation parameters on board, he called Julia on the phone and told her when to look out of the window. They waved to each other as The Station passed over Star City on a dark but clear night.

## To all those good people

I didn't put any acknowledgements at the front of this manuscript on purpose. Nobody ever reads those. For some reason, they are usually written very conservatively, no matter what the style of a book is. If you have read this far, I dare say you found the story interesting. This is one more little story I would like to share with you. It happened to me during the flight. This Book has happened to me. These people helped The Book to happen to me in one way or another. I am forever grateful.

Loredana Bessone. I often feel it's only by chance that we were born in this incarnation in different countries and from different parents. She's like a sister to me. I lived with her for part of the six months of Frank's mission when being alone was bearable no longer. I wrote part of this book on her home office desk. What else can I say? I could write a separate book just about me and her. She is a true blessing in my life.

Oddrun Uran and Donny Aminou are the only close friends living in my geographic neighbourhood (in Dutch terms, reachable by bicycle). They took it upon themselves to invite me for dinner at least once a week or any time I wanted company. Their gorgeous little Inna grew up and learned to talk in several languages while Frank was flying.

My friend Fatima El Amrani Jaafari, who is also my beautician. She didn't know who Frank was. I once went to her salon to refresh my permanent make-up:

"Your husband is away?" she asked.
"Right, how did you guess?"

"I know him, he doesn't like you doing things which need healing for a couple of days."

Indeed, she knows him but we never talked about his work. This is so refreshing. A couple of weeks later, I met Fatima again.

"I have the whole evening free," I said.

"Is Frank away again?"

"Yes, he's travelling a lot these days," I said, not lying. I just didn't want to talk about space. Everybody is talking to me about space. There are other interesting things in life.

"He has a tough job. Can't he change it?" she said.

"Not at the moment," I told the truth again. "Maybe in a couple of years, who knows. But now he really has to complete something he's doing.'

"OK, give him my best wishes when you talk to him."

I felt so guilty for not telling the whole story. Yet I knew that I didn't have to. Fatima is the most intelligent and cool person I know. Rooted equally in both European and Arabic cultures, not even 30 years old, she runs her own business as a real entrepreneur and yet remains understanding and caring as a real sensitive woman. Fatima, I'm sorry I didn't tell this to you before. It was too precious to me that we could talk about everything else in the world apart from my personal life and the extravagant whereabouts of my husband. It means a lot to me that you love me for who I am irrespective of who I am married to.

My friend Kim Appeltans has this amazing ability to surface in my life infrequently but on the most unexpected and meaningful occasions. This time she turned up right after I got home from the launch. I was dumbstruck by the static silence of being alone at home instead of being constantly on the road inseparably with Frank. I have never thought about it in those terms before but quickly discovered that such a sharp change of speed makes you feel dizzy in various ways. Kim dragged me to Boston with her for a few days. This gave me a nice buffer for readjustment to my new solitary reality. I needed that.

Everything is fine. Life is as good as you make it. I have huge room for creativity and imagination to fill my earthly space while Frank is away. I thought that I was doing rather well, until September for some reason turned out to be a very heavy month. Francois Caquelard and Alla Vein are the best doctors in the world, and fortunately they also happen to be our friends. Aquinia van de Zandt is another dear friend who happens to be the best alternative medicine practitioner. At different moments, the three of them carried me through some tough moments with their wisdom, kindness and confidence.

Frank's mom Jeanne and sister Carine were always there for me when I needed them. They saved me from the broken microwave when our house became unhappy in September. Together with our extended Belgian family, Gery, Marieke, Bart, Marleen, Wesley, Annick, Klaartje, Ine and Jef, they made these six months easier for Frank and me. We live in different countries and see each other only occasionally. But I know that I'm always welcome there and it makes all the difference.

Our Russian family, Lida, Lev, Alex, Tanya, Liza, Danya and Katya; Frank's kids, Tom, Nele and Koen, are all wonderful and important people in my life and I love them dearly. Nele came to spend a couple of days with me in Moscow before the rest of our Belgian family arrived. We had wonderful girly time together, and this helped us to ease the tension of the expectation of the landing. And we had a lot of fun when Tom, Koen, Jeanne, Carine and Jack joined us on the eve of the landing

Ronald, Ludmila, Alex and Philip van Hirtum adopted me into their family in the middle of October when I couldn't bear being alone any more. They gave me a home until I joined my family in Moscow prior Frank's return.

Julie Brown and Dixie Dansercoer introduced me to Davidsfonds publishers who took on my work in Belgium. This was an extremely exciting experience for me – my first ever publishing contract. Thank you all for your interest in my work, kindness, patience and support.

Our friend Denis Trussevich drew the wonderful sketches for the whole book, took my photograph and made a mock-up design for the cover. Thank you Denis, you have always been an inspiration for creativity.

AFTER VSTRECHA: FINALLY BACK TOGETHER

Carl Walker. He stopped me from trying to be too clever for my own good. He tolerated my arguments, while showing me where I could do better. In his selfless and committed 'hobby work', he not only gave me a lot of his free time (thanks to his partner for supporting this) for improving my texts, but also more in helping me find a path into the English language publishing world. I went onto Facebook to see Carl's profile when we first met. On my way out of that site, I hit a wrong button by accident and by doing so invited my entire email address book (over 600 people) to see my photographs (even thought I didn't actually have any in my profile). Not only did this invitation go out, but it kept sending reminders. After the fourth reminder, I was getting desperate. Even when I left Facebook, this didn't stop the reminders from going out. It took Loredana two hours of intense work to help me cancel Facebook so that it cut the links that were sending spam email in my name. But if this mess was the price for getting to know Carl, it was totally worth it! Now I am back on Facebook again with a page called Lena De Winne. My Countdown. If you are on Facebook, please "like it" and invite your friends. Spreading the word will hopefully increase sales and generate more money for UNICEF Belgium. (By the way, Frank is not on Facebook or on any other network. If you find his profile, don't believe it, somebody is unlawfully impersonating him. Unfortunately it seems that nothing can be done to stop this.) Through Carl Walker I met Robert Godwin who became my first English language publisher. This was a major step for me to be accepted outside Belgium where a story about Frank has a chance to be successful because of the public interest in his personality. Thank you Robert for your trust in me.

Caring and supportive Belgian people: Minister Sabine Laruelle, Monique Wagner Emmanuelle Courtheoux, General Jef Van den put, Professor Floris Wuyts, Professor Andre Aubert, UNICEF Belgium, the Belgian embassy in Moscow. They all made an effort to stay in touch and welcomed me while Frank was in space. It was an honour and a very special experience to be included into 'Belgian life' during these six months. Belgium is becoming my third home. Frank's friends became mine. Jack and Carmen Waldeyer; Jos, Andree and Julie van Schoenwinkel; "Barny" and Mady Flamang; Herman and Christiane Hendrickx; Luc Tytgat and Patricia Mommens together with Brice, Lea and their dog Lola. Everyone reaching out to me from Belgium helped me keep warm while The-Belgian-Of-My-Life was weightless and away. And Piet Porrey and Veerle van Hoof have also offered their very healthy and fruity PIPO products.

The 134 Polythech Promotion – Frank's alma mater. They included me in all their email chatter, invited me to their gatherings. I didn't have a chance to go but I'm looking forward to meeting them all one day.

My Welsh friend who likes going by the name Celt is a master of human communication and knows more than Wikipedia and Encyclopedia Britannica combined. Thank you Celt for helping me look for publishers and agents. Your depth and extravagance, combined with your admirable dedication to challenge me, kept me on my toes and gave a lot of encouragement.
I also would like to thank for the warmth of their attitude to me, in my life in general and during the year 2009 in particular: Mark Mouret, Hans Bolender, Arie Bossche, Silke Reukauf, Victor Nikolayev, Michel Praet, Beata Mirtchouk, Masha Scherbinina, Dzana and all the friends who are mentioned in this Book.

The community on the Russian 'Snob' web portal expressed a lot of unexpected support in response to my blog about Frank's flight. Despite the distance inherent in any virtual world, the genuine positive interest, encouragement and appreciation I felt within this little world, which was designed to connect global Russians, has contributed to my overall sense of confidence and self-esteem.

And of course The Crew – Bob, Roman and Frank, and The Wives – Julia and Brenda. This entire Book is my tribute to them, their courage and kindness. Dear friends, please accept my humble gratitude for sharing this magic experience of a tough adventure and incredible human connection. And Frank, hey, I love you!

## To be continued

Like almost every woman, and many men, I get to cry every now and then. It happens naturally when in physical pain, or occasionally because of anger or helplessness, but most often from some unexpected strong sentiment. I find it particularly touching when, in the movie 'Notting Hill', the heroine played by Julia Roberts accepts to stay indefinitely in England with the hero of Hugh Grant, not to mention Lassie finally coming home.

Three, two, one years before the launch, when Frank's flight seemed remote and unreal, my imagination drew every now and then some scary pictures that I couldn't fight off immediately. They overpowered me sometimes and made me cry. But this only happened rarely.

For some years I couldn't stand watching launches: I was terrified of something going wrong. I couldn't stand the idea of being in the position of a spouse who had to deal with this. And yet I knew that one day I would have to watch Frank's launch.

One day while we were in the Houston airport lounge waiting for our plane home, a Shuttle launch was being broadcast live on CNN. I had to run away and hide in the bathroom. NASA astronaut Janice Voss was commentating on the event live in the studio. In the voice of this otherwise very calm, correct, super lady, there were lights and flares, excitement and sparkles, and she was having a ball!

"I have to remember this feeling of overpowering happiness," I thought to myself as I was saving my nose from excessive swelling and eyes from total loss of make-up, and hoping that no one entered the bathroom. As always, Frank was sympathetic to the fact that I was upset but still couldn't quite get it – what was there to cry about? Men. "I need to call Janice and ask her where her source of happiness is," I thought, as I searched for a positive thought to concentrate on.

***

We are all middle aged. Only Roman and Julia are still below 40. And yet something amazing happened here, a miracle that doesn't happen when you are grown up with all the usual stresses and hustles of an adult life, which is always too short on time and too heavy on logistics. We became friends in a way that is more than what makes a friend in the adult

Western world – a person you happen to know and mean to invite for a drink if they're in town. We became kind of a family. We live on different continents and in different time zones. We have different priorities and different plans for after the mission. We will not be seeing much of each other anymore when post-flight tours are over.

Unfortunately this is the reality of our good adult lives. But like those military school friends of Roman, who just picked up and came from all over the world to Star City to have a drink with him on the evening of his official flight assignment, like Loredana, Ludmila, Oddrun, Tanya, Kate and Olga, who held me up with their love and care when I was collapsing, we will always mean to each other much more than just somebody we used to know in previous years of our lives.

One of Nietzsche's most famous aphorisms is that, "If it does not kill you, then it makes you stronger." The first and the last time I cried around the time of the flight itself was when I was saying goodbye to Julia, the day after the docking. I was going back home to Holland, she was staying in Star City, starting her new job (in the newly formed cosmonauts' training centre, which had just been turned from a military unit into a civilian organization). We made the previous two months much more bearable for each other. I reread the word 'bearable' I just wrote. Did I mean it? Maybe I did. Or maybe 'possible' would be more correct. We could help and support each other like no one could help and support us because we were sharing the same experience at the same moment in life. I'm sure that if we lived next door we would have become inseparable friends. I only had friends like this in my childhood. Tears still come to my eyes as I'm writing this now. But now I almost know how to stop them – half way. I keep repeating my silent prayer to the Universe to have the wisdom to use for best all this newly acquired strength.

Leiden, Star City, Cologne 2009

## ABOUT UNICEF :

Frank De Winne is a goodwill ambassador for UNICEF Belgium. Part of the profits from this book are contributed to UNICEF Belgium in support of the worldwide UNICEF-programs.

UNICEF was the first organization to make use of celebrities to promote its activities to the general public. Danny Kaye and Audrey Hepburn opened the path for many others to follow.

UNICEF Belgium can count on the support of well known persons to ask the general public for attention and support. In 2003 ESA astronaut Frank De Winne became a Goodwill Ambassador for UNICEF Belgium. This was the start of a very fruitful cooperation. In 2003 Frank travelled to Benin in Western Africa to visit projects for the prevention of child trafficking. In 2006 he travelled to Pakistan together with UNICEF Ambassador and polar explorer Alain Hubert to report on the consequences of the earthquake which had hit the country. In 2008 – while in full preparation for the OasISS mission – Frank made time to visit UNICEF projects on Water, Sanitation and Hygiene (WaSH) in Mali. This visit took place within the framework of the UNICEF Belgium "WaSH-campaign" for which Frank became patron. Before leaving for his second space mission, Frank and Lena officially launched the WaSH-campaign by breaking the world record for the longest toilet queue ever; together with 755 other participants.

While in space Frank De Winne did not forget about UNICEF. His blue T-shirt went with him on the mission and during his stay in the ISS he used every possible opportunity to strengthen UNICEF's message for the well-being of the world's children: from a greeting recorded for U2 which was broadcast during their 350° World Tour, over a radio contact with children in the Village of Gao in Mali to listening to the international UNICEF-Hymn in the ISS to celebrate the 20th anniversary of the UN Convention on the Rights of the Child.

The contrast between the high-tech world of space aviation and the poverty in many parts in the world is seen by Frank as a challenge rather than an inhibition. *"I am truly convinced that the technology used in the ISS will be able to give very concrete spin-off applications to improve the lives of children on earth. And UNICEF is one way of doing so"*, said Frank De Winne.

More information on UNICEF and the UNICEF Belgium WaSH-campaign: www.unicef.be

[1] In early 2010 (after the book was finished) the US administration has changed the approach and opened space transportation market to the commercial companies. The current options include a possibility to fly crews in Dragon vehicle on Falcon-9 launcher produced by SpaceX.

[2] The last crew rotation by a Shuttle was done at the end of 2009. After that all crew rotation is happening with the Russian Soyuz vehicle.

[3] The first Korean astronaut Yi So-yeon flew to space 8 – 19 April 2008. Ko San served as a back-up.

All rights reserved under article two of the Berne Copyright Convention (1971). No part of this book may be reproduced or transmitted in any form or by any means, electronic or mechanical, including photocopying, recording, or by any information storage and retrieval system without permission in writing from the publisher.

We acknowledge the financial support of the Government of Canada through the Book Publishing Industry Development Program for our publishing activities.

Published by Apogee Prime, A division of Griffin Media
http://www.apogeeprime.com
Printed and bound in Canada

My Countdown /Lena De Winne
First English Edition

© 2010 Apogee Prime/Lena De Winne  ISBN 978-1926837-09-3
Editing by Carl Walker
Layout based on the layout by Peer De Maeyer
Cover by Robert Godwin
Illustrations by Denis Trussevich
Photos courtesy NASA/ESA and the personal archives of the De Winne and Romanenko families, Picture of Guy Laliberté courtesy of Star City/Andrey Schelepin